# First to Fly

# First

**Thomas C. Parramore**

*North Carolina and the Beginnings of Aviation*

# to Fly

*The University of North Carolina Press  Chapel Hill & London*

The North Carolina Department of Cultural
Resources and the First Flight Centennial Commission
generously supported the publication of this work.

Designed by Richard Hendel
Set in Cycles, Poplar, and Meta Black by Eric M. Brooks

The paper in this book meets the guidelines for
permanence and durability of the Committee on
Production Guidelines for Book Longevity of the
Council on Library Resources.

Library of Congress Cataloging-in-Publication Data
Parramore, Thomas C.
First to fly: North Carolina and the beginnings of
aviation / Thomas C. Parramore.
   p. cm.
Includes bibliographical references and index.
ISBN 0-8078-2676-6 (alk. paper)
1. Aeronautics — North Carolina — History.  I. Title.
TL522.N8 P37   2002
629.13'09756 — dc21       2001049162

06  05  04  03  02   5  4  3  2  1

*To my wife and soulmate, Barbara*

# contents

# maps & illustrations

I was educated, from start to finish, in the North Carolina Public School and University Systems but did not learn that this state had an important aviation legacy reaching back over a century before the Wright brothers and decades after the first powered flight. Traditional history texts do not include this topic. Railroads, ships, automobiles, even rockets are covered; yet airplanes, helicopters, and dirigibles are not.

Eventually, I stumbled, quite by accident, on this buried aviation legacy and understood why it remained unexplored by historians. It was because the story is found only in bits and pieces in thousands of old (and some not-so-old) newspapers, treasured personal papers of many families, and a thin miscellany of mostly timeworn books and magazines. Even early twentieth-century North Carolinians, unable to glimpse the picture as it related beyond their own bailiwicks, failed to appreciate that a legion of Tar Heel artisans built airplanes, flying boats, helicopters, and dirigibles from at least 1873 to 1923. During this half century, Tar Heels launched over three dozen aircraft projects. Not all resulted in a completed flying machine, still less ones that actually flew, though some did.

The story revealed in these scattered fragments is that North Carolina artisans labored at or near the cutting edge of aviation technology and that the state produced some of the most notable airmen and women, including professional balloonists and parachutists, of the half century that encompassed the work of Wilbur and Orville Wright. Many of these fliers set aviation records in their fields or were the first to perform notable exploits. Perhaps this volume will help bring that story to the attention of historians and teachers who can then fill the gap in our history books and classrooms.

There is another audience that should also be exposed to our aviation legacy. At this writing, we approach the centennial of the first airplane flight. But the hordes of visitors it will draw here may recognize North Carolina only for having provided a patch of sand, in a remote corner of the state, for that fateful event. Will they come and leave with no understanding that the first powered flight falls within the context of a vital half century of Tar Heel aviation? Will they leave unaware that scores of Tar Heels aided the Wrights in their work, that it is unlikely the first powered flight would have occurred as early as 1903, and probably not in America, but for these North Carolinians?

In that case, the celebration—here and elsewhere—of the first powered flight may not recognize, in any meaningful sense, North Carolina's aviation legacy; it may be seen as alien to our state's history and the concerns of its citizens. "First in Flight" would then be confirmed as the hollow boast many have said it was all along. Maybe we can prevent this from happening.

Few books are the work of a single person, and this one is no exception. Over the past several years, I have benefited from the aid and encouragement of many people. Among these are the staffs of the University of North Carolina (UNC) at Chapel Hill's North Carolina Collection; North Carolina State University Archives, Raleigh; Wake Forest University's Baptist Collection, Winston-Salem; National Air and Space Museum Archives, Smithsonian Institution, Washington, D.C.; Library of Congress, Washington, D.C.; Wright State University Archives, Dayton, Ohio; Norfolk, Virginia, Public Library; Manteo, Asheville, and Clayton, North Carolina, Public Libraries; North Carolina Division of Archives and History, Raleigh; and Cape Fear Museum.

Many individuals have allowed me access to their personal and family files. They include Grayden M. Paul, Professor Charles Paul, Lina Bilar, and E. K. Liberatore (for William Luther Paul); Gloria E. Sodergren (James S. Spainhour); Thomas A. Allison (Raymond Vance Allison); Charles E. Moore (Hubert McLean); Caralee Maynard Rooks, Carolyn G. Blackstock, and Sylvia Maynard O'Kelly (Belvin Womble Maynard); Hal and Dave Stewart (Alton Stewart); John Foster West (Charles W. James); Gerhart Everwyn (David Palmgren); John R. Bellefleur, Wesley R. Smith, George Stevenson, Sister Virginia Geiger, and James Donelson Christmas (William W. Christmas); Margaret Kromm Federline and Donald Kromm (Louis Henry Kromm); Willia Severn Raye (Bennett D. Severn and Irene Tate Severn); Carr Booker Jr., Roy L. Booker, Robert A. Booker, J. W. Fuller, and Karen Booker (Carr E. Booker); James E. Meyer (Thornwell H. Andrews); Glenn L. Moore (Harold Chase and M. F. H. Gouverneur); Dr. Mark Bensen (Igor Bensen); Ellenore Eddy Smith (Lincoln Beachey); Hardy Mills (Augustus Leazar); Reverend W. Stephen King (Jacob Aaron Hill); Mrs. Melba Ambrose (John Maynard Smith); Daniel Grady Tate, Bill Harris, and Al McSurely (Wilbur and Orville Wright); Anna Pennington (Warren K. Pennington); Harry Thompson (William Henry Rhodes); Sue Rouse (Kinston Demoiselle); and Bill Bailey (Polk Denmark). Tom D.

Crouch of the National Air and Space Museum and aviation author William F. Trimble kindly read the manuscript and offered valuable suggestions.

For the opportunity to publish this book, I am indebted to Betty Ray McCain, Secretary of Cultural Resources; David Perry, editor-in-chief of UNC Press; and Kathryn Holton, director of the First Flight Centennial Commission. For assistance in various forms, I thank especially Leo E. Opdycke and Mitchell Bowden. Finally, I offer my thanks and apologies for those omitted from this list.

# First to Fly

*If there is a domineering, tyrant*

*thought, it is . . . that the problem of*

*flight may be solved by man. Once*

*this idea has invaded the brain it*

*possesses it. . . . It is then a haunting*

*thought . . . , impossible to cast off.*

*—Louis Pierre Mouillard, 1881,*

*quoted in Walsh,* One Day at

Kitty Hawk, *1975*

# To Snatch the Secret
## from the Womb of Time

THE CONJURER'S DECREE

No grander quest has inspired mankind, no greater single achievement blessed its efforts, than the ability of humans to fly. The twentieth century, for all its cruelty and suffering, stands forth in the annals of history for its conquest of powered flight, the glory, indeed, of the Second Millennium. Well before the century ended, flight rewove the fabric of human life, altered our perceptions of time and space, and galvanized and exalted our faith and self-confidence in what is to come.

For three decades before and two more after 1903, North Carolina could, but did not, claim to stand at the forefront of aviation's advance. The reason no such claim was made is that the state did not realize that it was entitled to do so. The evidence for these contentions comes in many forms and was signified two centuries before the first flight was performed.

Explorer John Lawson traveled to Carolina mainly to glean information useful to the colony's English proprietors. The splendid book he published in 1709, *A New Voyage to Carolina*, revealed a treasure of data on the settlement's climate, geography, soils, minerals, flora, fauna, and inhabitants. It was especially interesting for its intimate glimpses of Carolina's aborigines, their customs and beliefs, arts and crafts, and patterns of living.[1]

Around 1708 Lawson visited several Native American tribes of the Albemarle Sound area. Although a thorough rationalist by training and predilection, he was charmed by stories of folkways and the occult among these Native Americans and passed on many representative tales. Clients of Tuscarora tradesmen, he wrote, bought liquor by the mouthful, usually with the collaboration of the man with the largest mouth; the exceptional marksmanship of the Flatheads came from strapping their infants' heads to boards so tightly that their eyes bulged, making them splendid hunters.[2]

An extraordinary tale of the supernatural concerned the celebrated "great Conjurer," a Chowan Indian named Roncommock, perhaps still alive in 1709. "Persons that affirm they were Eye-Witnesses," wrote Lawson, insisted that they had seen Roncommock "take a Reed about two Foot long in his Mouth, and stand by a Creek-side, where he call'd twice or thrice with the Reed . . .; and, at last, has open'd his Arms, and fled over [Salmon] Creek, which might be near a quarter of a Mile wide or more."[3]

This wonderful feat, deemed "seemingly true" by the worldly Lawson, was allegedly performed on the plantation of former governor Seth Sothel, himself a proprietor of the Carolina colony, at the mouth of Salmon Creek in Bertie County. It was also the site of the first English house built in North Carolina. Bertie happens to be the birthplace of a poet whose verse was among the first, if not *the* first, to celebrate the advent of powered flight in America. Forty miles north of Salmon Creek, North Carolina's first airplane was built and tested. Eighty miles southeast, humankind's first helicopter lifted vertically off the ground.

The birthplace of an aviator who built and flew a biplane and, a quarter century later, patented an enormous flying-wing airliner is eighty miles northwest; that of a pioneer of the automatic pilot in planes and rockets, thirty-eight miles northwest. Seventy miles north, Norfolk, Virginia, witnessed the first flight from a ship's deck. And at a barren sand dune sixty miles due east of Salmon Creek, a balding young man, on December 17, 1903, flew a motored stick-and-rag biplane for twelve seconds.

***The Flyer,*** **Theodor De Bry's engraving of John White's watercolor of an Indian conjurer, 1585, on the coast of present-day North Carolina. North Carolina Collection, University of North Carolina Library at Chapel Hill.**

"Witnesses" to Roncommock's flights harbored no doubt that humans could fly—if they possessed the Conjurer's secret. How to acquire or fabricate that power was what French gliderman Louis Pierre Mouillard termed the "tyrant thought" that ruled over humankind for two centuries after Roncommock's time and still impels us relentlessly toward the infinite. The magical reed can be said to symbolize a quest that began long before the Conjurer's day and continues today. It is nothing less than our emergent hope of navigating the solar system as readily as we do the paths and sea-lanes of our native planet. Roncommock's reed challenges us to burst Earth's coils and soar—in body as well as spirit—to the outermost reaches of the universe, the innermost depths of our being. The site where the Conjurer flew is the epicenter of a region where humankind was baptized in the transcendent experience of both horizontal and vertical powered flight.

The ability of humans to fly, a goal finally achieved in North Carolina, summoned forth, between 1909 and the end of World War I, the greatest surge of mechanical creativity and individual achievement in the state's and the nation's history. The artisans who aspired to fly and the warriors who set out in 1914 to meet enemies in an utterly new medium of battle shared a common spirit of courage and experimentation.

The tale that follows deals with humankind's earliest attempts to break the bonds of gravity, to master resisting winds and subdue foreboding skies. It is a story that began, in its practical phase, longer before than since the flights at Kill Devil Hill. North Carolina was "first in flight" (as its license plates boldly declare) well before the Wright brothers' success in Kitty Hawk confirmed what was ordained by Roncommock's mysterious reed.

Ancient legend teems with exploits of flying humans. That a Daedalus or an Icarus tried to escape on waxen wings from Minoan Crete and its fabled labyrinth is, like a hundred such tales, a myth. But sound authority holds that Archytas of Tarentum, in fourth-century-B.C. Greece, experimented with kites and toy flying machines, including a wooden dove that flew by "the secret blowing of air enclosed inside," perhaps a primitive compressed-air mechanism.[4]

Attempts to fly are impressively recorded in Medieval and Renaissance literature. They include a flight with artificial wings for a distance of "125 paces" by an eleventh-century English monk, Oliver of Malmesbury. A Saracen of twelfth-century Constantinople crashed after skydiving with counterfeit wings from the top of a tower. Leonardo Da Vinci sketched numerous flying machines in the fifteenth century, and in the early sixteenth an Italian is said to have broken a leg trying to fly on cock-feather wings from Scotland to France.[5] These men did not believe that they were on a fool's errand. If humans could walk like apes and swim like fish, they could learn to fly like birds. Roncommock may have foretold with his reed that the vault of his secret would someday be unlocked—in his own bailiwick—and swing open for all humanity.

Meanwhile, imagination ventured boldly where reality dared yet tread. Englishman Francis Godwin in 1641 spun a tale, *The Man in the Moon*, of a Spaniard borne aloft in a car drawn by swans. Another, Samuel Blunt, in 1727 wrote his whimsical *Voyage to Cacklogallinia*, in which an aeronaut's car, pulled by roosters, flies (like Godwin's hero for chicken feed) from Jamaica to various countries. Tales of this sort grew more realistic as experimenters made the first tentative progress toward mastering the art of flight. *The Discovery of the South by a Flying Man*, by Frenchman Restif de La Bretonne in the 1770s, thus bears its hero on mechanical wings worked by springs. Once flying was understood to be a feasible technological challenge, solutions were not far off.[6]

After 1783, when French balloonists succeeded in breaking earth's bonds, there was no need to limit tales of human flight to mere fancy. A German writer, in 1789, wrote what seems to have been the first novel founded on the new reality of human flight. His name was Johann W. A. Schofel, and his tale was entitled *Hirum-Harum: Ein Satirsch-komischer Original-Roman* (A Satiric-Comic Novel).

In company with pioneer French balloonist Jean-Pierre Blanchard, the central character builds a balloon in which he flies alone from Paris to North America. After many adventures, he crashes just outside the village of Salem (now Winston-Salem) and is conveniently rescued by German-speaking Moravian settlers. Schofel even cites a fictional Salem printer of that era as his book's publisher.[7]

CAXTON'S ODE

Powered flight, like manned flight, began in France. In 1852 Henri Giffard briefly flew his steam-driven dirigible (i.e., steerable) balloon over Paris. Another flight with such a machine was made in California in 1869; the *Avitor Hermes Jr.* was the work of transplanted Englishman Frederick Marriott. (Hermes was the Greek God of astronomy, avitor a form of the term "aviator.") On July 2, 1869, it made an unmanned flight of a mile at San Jose's Shellmount Park racetrack, the first powered flight in America.[8]

Among the invited spectators that day was Judge William Henry Rhodes, a native of Windsor, in Bertie County, who had written poems and stories since childhood and then exploited his skill in the West.[9] Two of his poems, both published in anticipation of the dirigible's flight, were among the earliest American celebrations in verse

The frontispiece sketch in Johann Schofel's 1789 novel, *Hirum-Harum: Ein Satirsch-komischer Original-Roman*, depicts a fictitious transatlantic balloonist apparently being rescued by residents of Salem. North Carolina Collection, University of North Carolina Library at Chapel Hill.

of Marriott's work, perhaps even the first inspired by the new reality of powered flight.

Rhodes was born in Bertie in 1822 and received his higher education at Princeton and Harvard Colleges. After graduating in law from Harvard, he opened a practice at Windsor, near the head of Roncommock's Salmon Creek. When his father, Colonel Elisha A. Rhodes, was named U.S. consul to the Republic of Texas in 1838, the son joined him there before following the Gold Rush to California. Prior to leaving Texas, Rhodes won literary notice with the publication in 1846 of a small volume entitled *The Indian Gallows, and other Poems*. Its selections included legends of Indians, pirates, and others from his North Carolina past. In California, where he became a probate judge and gubernatorial private secretary, Rhodes earned acclaim with his short stories, written under the pen name "Caxton." He was especially admired for his science fiction, including tales of flying experiments and adventures, a genre in which he became a recognized pioneer.[10]

Marriott's machine was an extension of his earlier work as an asso-

ciate of the gifted English aerial experimenters John Stringfellow and William Henson. *Avitor Hermes Jr.* was a thirty-seven-foot hydrogen-filled airship with wings of cloth-covered wire, a steam engine, rear elevator, and rudder.[11] Rhodes first honored Marriott's work in a poem published in a San Francisco newspaper in early 1867. Entitled "The Avitor," it heralds the flying machine not simply as a revolutionary mode of travel but, astutely, as a means to emancipate the human psyche, a theme destined to appeal to generations of poets to come. The future of flight, he predicts, will be grand beyond any of humankind's previous achievements:

> Hurrah for the wings that never tire—
> For the nerves that never quail;
> For the heart that beats in a bosom of fire—
> For the lungs whose cast-iron lobes respire
> Where the eagle's breath would fail.

Airships, the poet foresees, will speed immigration to the American West:

Sierra Nevada's shroud of snow
And Utah's desert of sand
Shall never again turn backward the flow
Of that human tide that may come and go
To the vales of the sunset land . . .

And, he predicts, airships will foster geographic discovery:

No longer shall earth with her secrets beguile,
For I, with undazzled eyes,
Will trace to their sources the Niger and Nile,
And stand without dread on the boreal isle,
The Colon of the skies![12]

Rhodes's clairvoyance is shown to even better advantage in lines
he penned and published in March 1867 to hail Marriott's unfinished
*Hermes*, a second dirigible. Here, he anticipates that aircraft will
someday become the means for scientific probing of the extent and
nature of the universe:

An ocean rolls, whose billows never rest,
Whose depth no plummet sounds, whose width no eye,
The landless, shoreless ocean of the sky.
Up on those waves see 'Hermes' climb,
To snatch the secret from the womb of time!

That electronic planets would one day circle the earth, bearing
male and female crews of multiple nations and races, is testimony to
Rhodes's foresight:

Henceforth our race are brothers here below,
Henceforth fraternal blood shall cease to flow;
No hate shall thrive upon a mountain chain,
For mountains now are level with the plain; . . .
No quarrels mingle with the ocean's roar;
THE HERMES FLOATS, AND OCEANS ARE NO MORE.[13]

Rhodes died in San Francisco in 1876, leaving a widow and seven
children. In the twentieth century he has received belated recognition as a midwife of science fiction but not yet the credit he deserves

**Marriott's dirigible *Avitor*, which in 1869 performed the first powered flight in America at San Jose, California.**

as a prophet of flight, space travel, and their meaning for human-kind. A true herald of the sorcerer, he acknowledged in another poem "[t]he tawny native of the Indian clime / First in the Race of Science and of time."[14]

In North Carolina, as early as the 1870s, Frederick Proctor, an Elizabeth City eccentric, was noted locally for predicting horseless carriages, undersea boats, and "ships . . . that sail in the air." In 1902 journalist W. O. Saunders recalled Proctor's exotic expectations and noted that, all those things having come to pass, "people who have not forgotten him feel repentant now that they should have treated his talk with so little credence."[15] Elizabeth City is only forty miles from the spot where Proctor's boldest dream came true.

Robert Frost was a lovelorn New Englander when, in 1894, he visited the fabled Outer Banks and experienced, on Kitty Hawk beach, the seductive symphony of wind, waves, waterfowl, and blowing sand. In his poem "Kitty Hawk, O Kitty Hawk" (1953), he recalls:

Little I imagined
Men would treat this sky
Someday to a pageant
Like a thousand birds.

But planes and spaceflight became a leading theme of his poetry. At President John F. Kennedy's inauguration in 1961, Frost, as the nation's poet laureate, challenged the new president to follow in the steps of Wilbur and Orville Wright "as models of excellence." Two months later Kennedy convened a special session of Congress to seek a greatly enhanced space program. On July 20, 1969, Astronaut Neil Armstrong stepped onto the surface of the moon, bearing with him a piece of wing cloth from the 1903 *Wright Flyer*.[16]

An accumulation of portents and predictions, then, pointed, long before 1903, an arrow from Roncommock's quiver toward Kitty Hawk as the venue for the great transformation. From Kitty Hawk beach, Frost distilled the inspiration that may, at a critical moment, have rallied a tepid American space program, leading to the first moon landing.[17]

## THE GIFT OF THE MARTIANS

By the closing years of the nineteenth century, reports of experimenters actually flying their machines, or making promising attempts to do so, were so frequent that the public could be excused for supposing that practical flight was near at hand. Dirigibles, especially, were having limited success in overcoming the conundrums of aerial steering and stability. The federal government invested $50,000 in heavier-than-air experiments conducted on the Potomac River by Smithsonian Institution secretary Samuel Pierpont Langley. That ordinary people were ripe for stunning success in aviation, however much some pundits might deny the possibility, was revealed in the 1896–97 excitement over unidentified flying objects (UFOS), America's first such mass phenomenon.[18]

Beginning in California in the fall of 1896, thousands of people from coast to coast reported seeing cigar-shaped or cylindrical flying objects, some with airscrews or propellers, flashing colored lights and emitting music, hissing sounds, voices, and so on as they passed. The evidence was too widespread to be lightly dismissed. Six beings were said to have emerged from such a craft in Illinois and offered a ride to an earthling, who, perhaps to our grievous loss, declined. In

Arkansas, a state senator named Smith reportedly drew up a bill to levy a tariff on a UFO "flying through the air of the great state of Arkansas, . . . without paying taxes." Populist governor John Leedy of Kansas hoped that the monopolistic tyranny of railroads might be undercut by the new floating highways in the sky.[19]

A photograph showed an airborne flying machine near Chicago, but, regrettably, it proved to be deftly painted on canvas. Joseph Joslin of St. Louis was allegedly seized by creatures from a UFO and held captive for three weeks, though evidently gaining no enlightenment from his experience. On April 17, 1897, an airship reportedly crashed into an Aurora, Texas, windmill, killing its unknown pilot, whose corpse was interred at a public funeral! A resident of Washington State claimed to have started the furor by releasing a pelican with a Japanese lantern tied to its leg.[20]

This echo from Roncommock's reed began to stir North Carolinians in April 1897. Tar Heels, less hurried than most Americans, at first reported no UFO sightings. But a disposition to do so had manifested itself long before. In mid-1806 settlers near the base of a mountain in Rutherford County testified before public officials to seeing "a thousand or ten thousand things" in human form "flying in the air" on the mountain's heights. Closer inspection revealed "a very numerous crowd of beings resembling the human species" but lacking "any particular member of the human body, nor distinction of sexes." They were said to have gathered at the Chimney Rock, a 225-foot monolith emerging from a shoulder of the mountain. Clad in "brilliant white raiment," they were seen to rise in pairs and trios to the top of the rock and then vanish, leaving, according to witnesses, "a solemn and pleasing reflection on the mind, accompanied with a diminution of bodily strength."[21]

North Carolina's turn to observe the latest aerial visitors came on April 5, 1897, when people in Wilmington squinted at a "brilliant floating mass in the heavens to the west of the city." The intruder, advancing rapidly upriver from the direction of the Atlantic Ocean, passed the city opposite the Market Street dock. Its course, for reasons known only to its spectral self, took it "in the direction of the Navassa guano works."[22]

The next day a Wilmington paper cited "gentlemen who saw the ship through field glasses" and discerned "wires and ropes about it." Others saw colored lights, and "some of our . . . most reliable citizens" saw a "network about the aerial wonder." A few heard "the tingling of bells on the air ship." The visitor then sped off toward some celestial rendezvous. A Wilmington editor, ridiculed for his credulity by some who had not seen the airship, asserted that he had watched it himself and mused whether it might be an omen of a coming calamity. If weathermen supposed it was only a balloon, then how could it move across the sky "with such speed"?[23]

The *Wilmington Star* on April 8 dismissed the airship as a hallucination, witnesses notwithstanding. The UFO had last been observed "scooting over Lumberton [eighty-five miles northwest], Monday night." Half a dozen Lumbertonians saw a "brilliant electric light" moving southwest and darting up and down, disappearing for a few minutes and then reappearing. On April 11 and 12, in and around Wilmington, there were more sightings of a "large light . . . sinking rapidly towards the west." Some thought it might be "the evening star," but "its sudden movements" remained inexplicable.[24]

At the town of Wilson, a bright beam, "about the size of a large incandescent light," appeared to the northwest around 8:30 P.M. on April 11. It descended toward the earth, rose again, sometimes nearly extinguished, then brighter. Nash Street witnesses thought it might be "Edison's electric star that he had sent up in Pennsylvania," "a balloon," or even "a sure enough 'air ship.'" On the night of April 8 citizens of Fayetteville, one hundred miles northwest of Wilmington, had spied "a big ball of fire about the size of a football field pass directly over" the town. The light seemed to be "the searchlight of the air ship." It passed west of Fayetteville again about 8:00 P.M. on April 15. Some observers saw rigging and red and green lights.[25]

On April 13 residents of Williamston, 140 miles northeast, were startled by "a black mass against the moonlit sky, going slowly from South to North" and becoming "brilliantly illuminated." Its altitude was "not great," so its rigging could be seen by the naked eye. It ascended rapidly to a great height, throwing "a brilliant light . . . far ahead of it," and vanished.[26] Twenty miles south of Chadbourn (due west of Wilmington), it had been seen by "a reliable countryman" around 2:00 A.M. on April 7. Its powerful light for five minutes rendered "objects as distinct as day" before "it shifted and disappeared." At Kenly, in Johnston County, the mayor, chief of police, a prominent physician, the railroad agent, and others spotted it around 7:30 P.M. on April 22, "very near the earth. Its sails or webbing were plainly visible and resembled mosquito netting. . . . [T]he outlines of two persons were plainly seen among the netting." It moved away, to the southwest, and was seen no more.[27]

The westernmost sighting was at Chapel Hill in mid-April. Some students at the University of North Carolina, walking at night in the suburbs, observed "the winged wonder." Its "flashlight" was lit but soon vanished, then reappeared "with a dazzling brilliancy." Moving "as noiseless as . . . some terrible spectre," it sped away into the Orange County night. A final sighting occurred on the night of May 3. The *Lumberton Robesonian* related that A. W. McLean and S. B. Lewis noticed it floating "smoothly and slowly" southwest. (A barbell-shaped UFO-like object, partly excavated in 1917 at Delco, twenty miles west of Wilmington, seems to have swooshed away on its own, disappearing before its excavation could be completed.)[28]

Some Tar Heels who only read about the UFO dismissed it as a meteorological phenomenon. The editor of the *Goldsboro Headlight* called it "The April Fool Airship." He drew malign satisfaction from a report that residents of Galesburg, Michigan, had seen an airship explode in the sky, a fitting end to the "airy romance with which jokers have been . . . exciting credulous persons. . . . Like novelists who get their heroes into desperate situations from which there is no escape but suicide, the constructors . . . of imaginary airships have been compelled to extricate their machine from the . . . embarrassments which encompassed it."[29]

Many remained intrigued. A Wilmingtonian supposed that the airships were "of other planets . . .; the aerial navigators were out in space to see what they could discover." Maybe "the inhabitants of Mars or another planet are ahead of us in producing air ships."[30] More typical was the suggestion of the *Wilmington Messenger* that the visitor might be "an aerial ship sent up as an experiment by some ingenious inventor." One often read of just such craft being tested by American and European builders. The *Chadbourn Journal*, alarmed by the military implications of the high-tech vehicle, warned that, if "it was a sure enough air ship . . . in less than six months Uncle Sam's navy won't be worth old iron." The *Sanford Central Express* sagely prophesied that, if it had been an airship, "they won't even look up at such a thing a hundred years from now. Mark what we tell you." Comets and meteor showers were not mentioned.[31]

The sightings of 1896 and 1897 exemplified the penetration of the popular mind by recent published accounts of experiments with dirigibles and heavier-than-air craft. Many of the sightings described machines virtually identical with those being tested by aeronauts on two continents. Nationally, the accounts tended to coincide as to the appearance and behavior of the airships. Just as UFO sightings after World War II foreshadowed the coming of space flight, those of the 1890s suggested that the public was unconsciously setting the stage for powered flight by willing such craft into existence. If so, it would soon succeed.

One may suppose that the wave of UFO sightings gave added impetus to the work of inventors such as Langley and the Wright brothers and hastened the day when the last obstacles to practical air travel must give way. A public insistent that it had already seen swift and highly maneuverable dirigibles fly stood ready to applaud inventors who offered them the benefits of such machines. On balance, Tar Heels appear to have accepted the apparitions as evidence of one or more man-made airships with amazing aerodynamic capabilities. Steam and gasoline engines, it seemed, might soon achieve for them what Roncommock had long ago demonstrated with his reed.

Popular imagination was barely ahead of the game. Before the echoes of the cigar-shaped UFOs of 1897 subsided, Alberto Santos-Dumont flew from outside Paris in such an object, circled the Eiffel Tower, and returned to his starting point without untoward incident for the first successful dirigible flight.[32] Few seemed particularly surprised.

In the *Charlotte Observer*, edited by Joseph P. Caldwell, North Carolina had the nation's leading press advocate of the coming of heavier-than-air flight. The paper was convinced of the cause by Samuel Langley's launch of an unmanned steam-driven model airplane over the Potomac in 1896, and it never relinquished the faith. Through frustrating years of failure by Langley and others, its editorial page rang often with the tag line: "We Will Yet Fly!" The *Observer* seems to have been the only American periodical to champion powered flight as editorial policy and well before it was achieved.[33]

Caldwell hailed Santos-Dumont in 1901 for flying his dirigible around the Eiffel Tower. But the limitations of dirigible and balloon travel led the editor to turn again and again to heavier-than-air experimenters for the resolution of the "tyrant thought." Caldwell lauded attempts by Alexander Graham Bell, Hiram Maxim, and others, but it was from Langley that he expected ultimate success. (A couple of Ohioans on the Outer Banks had so far neither sought nor gained public recognition, even in North Carolina.) What the paper anticipated was no balloon, "but a machine which flies of its own efforts and not because it is too light to stay on the ground."[34]

The *Charlotte Observer*, Caldwell wrote in July 1903, "has all along maintained that" Langley "is the only real inventor of the flying machine. He has been working on it for almost twelve years past" and "has improved his machine. . . . It is a flying machine in every sense of the word. It will not rise like a balloon but like a bird." Whenever his paper repeated its prophesy that "we will yet fly," it had Langley in mind.[35]

This attitude elicited the scorn of the *Norfolk Virginian-Pilot*, and in the spring of 1903 the two papers engaged in a windy polemic over the feasibility of heavier-than-air flight. Caldwell excited the Irish wrath of the *Pilot*'s Michael Glennan by commenting favorably in March 1903 on claims by a pair of steam-airship builders in pastoral Stokes County, North Carolina.[36]

On May 10 the *Pilot* replied in an editorial, entitled "The Airship Folly," by attempting to show the futility of such efforts. "Despite continued failure to achieve practical results," Glennan asserted, "and the financial and physical disasters attendant upon the attempt to navigate the air, inventors are still plugging away at the problem. Hope springs eternal in the breast of the would-be emulator of the bird." These experiments, the editor noted, had cost several lives. The laws of nature, after all, showed that the goal could not be reached. In the first place, air was an "attenuated" medium for travel, requiring a great deal of power exerted against it to achieve even a slight effect. Hence, efforts to make machines fly without a balloon were pointless; a heavier-than-air vehicle would expend most of its energy in steering and turning propellers. "The kick of a propeller against an elastic and intangible medium" would produce little or no forward motion.[37]

Such machines, Glennan conceded, might have merit as carnival attractions, but they could never serve useful purposes. Common sense reinforced science. An advantage of rail travel was that "the whole power of its locomotive can be devoted to drawing it. None is required to sustain, or steer it. . . . The locomotive has merely to overcome the inertia of the train and the resistance of the air." Its weight "is sustained" on a "smooth surface and moved by power applied to an unyielding medium." Like a sprinter against his blocks, a train's force was exerted against something solid. Heavier-than-air flying, though "an interesting fad," had "not the slightest probability that it will be anything more." Besides, even if one could travel this way, the chance of escaping with one's life from an accident was almost nil.[38]

The *Observer* failed to profit from the *Pilot*'s lessons in science and common sense. Caldwell was "mortified on account of such ignorance" as Glennan's but had few resources with which to counter it. In reply to the *Pilot*'s train analogy, he drew on Daniel Webster's view in olden days that railroads themselves were a folly. A train speeding at, say, twelve miles an hour, Webster had declared, would have too much momentum to stop at an intended station; moreover, its steam would freeze in winter and leave the vehicle useless until warm weather returned. The *Pilot* editor countered that he would be happy to try out "the Stokes county patent" as soon as he heard that Caldwell had survived a flight in it.[39]

So ended the debate in an unproductive standoff. It was ironic that seven months later the *Virginian-Pilot* would sing its own praises as the first newspaper in the world to announce that humans had conquered the sky—in North Carolina. On the very next day, however, the *Pilot* carried a comment that seemed to question the primacy of the Wright brothers' achievement. Two Stokes County brothers named Hill, the paper declared, had preceded the Wrights in attaining powered flight![40] Was North Carolina "First in Flight" *before* the Wright brothers? The answer is yes, in the realms of both fiction and fact.

## MASTER OF THE WORLD

Novelists, uninhibited by the constraints of either science or common sense, enjoyed more freedom than journalists to speculate. Jules Verne, who was self-educated in aeronautics and other sciences, achieved in his novels and stories a skillful blend of fantasy and scientific understanding. In 1904 he published his novel, *Master of the World*, about the mad scientist Robur, a Captain Nemo of the clouds, who invents a vehicle—the *Terror*—that travels by land, sea, air, and underwater. Robur intends to use his machine to conquer the world.[41]

Robur was a figure Verne had used in a tale two decades before as commander of a helicopterlike craft of Robur's own invention. So "Robur" was not, as might be supposed, a conflation of the names Orville and Wilbur, whose exploits burst upon the world just before the new novel was published in 1904. Nevertheless, *Master of the World* has a pronounced North Carolina venue, for Robur builds his machine in the bowels of Rumbling Bald Mountain, on the border of Rutherford and McDowell Counties, quite near spirit-haunted Chimney Rock in the southwestern part of the state. The mysterious noises that gave the mountain its name were evidently caused by a series of earthquakes that shook it in early 1874. Verne tucked this information away for later use and made Rumbling Bald the site of Robur's latest project. The mountain's noises were attributed to his forges and experiments as he assembled his infernal machine.[42]

In Verne's tale, fumes and soot from the mountain drift eastward

The *Terror* machine at its home base at Rumbling Bald Mountain, Rutherford County. From Jules Verne, *Master of the World* [1904], 176, North Carolina Collection, University of North Carolina Library at Chapel Hill.

across North Carolina, raising speculation and concern as to their cause. It is reported that fear prevails in Morganton and Pleasant Garden, the towns nearest the mountain, where a volcanic eruption is feared. A manned balloon is sent from Morganton to inspect the mountain, but winds drive it eastward to a landing near Raleigh. Federal investigators search the mountain but find nothing suspicious. Meanwhile, reports emerge of an "extraordinary vehicle" roaring along Pennsylvania roads at "tremendous speed," then news of a submarine off the New England coast, a high-speed boat, and a flying machine. All are discovered to be a single vehicle. Inquiry shows that the builder is Robur, that his craft is known as the *Terror*.[43]

An agitated Congress appropriates $20 million to buy the machine from its owner. But Robur sends a letter from his mountain to newspapers, scornfully rejecting the offer and claiming that he intends to rule the world, "and there is no human power capable of withstanding it." He returns periodically to Rumbling Bald, his aerie and supply base, to prepare his assaults. In a race over the Gulf of Mexico, however, frantically pursued by the navy, the *Terror* is destroyed by a bolt of lightning.[44]

**The destruction of the *Terror* by lightning over the Gulf of Mexico. From Jules Verne, *Master of the World* [1904], 204, North Carolina Collection, University of North Carolina Library at Chapel Hill.**

*Master of the World*, published a few months after the Wrights' first successes, was the first novel inspired by the reality of powered, manned, heavier-than-air flight. One can concede that it was coincidental that Roncommock sounded his reed a canoe ride due west across Albemarle Sound from Kill Devil Hill. And it was only happenstance that humankind's first fictional balloonist landed in the village of Salem, that a son of Roncommock's neighborhood wrote perhaps the first American poetry inspired by powered flight, that Bald Mountain spawned America's first fictional airplane flight.

It was also chance, of course, that a North Carolina newspaper first championed heavier-than-air powered flight as editorial policy; it was irrelevant that a resident of the town that served as the Wrights' commissary predicted powered flight in the 1870s. It was quite immaterial that the first four manned airplane flights were performed forty miles southeast of there in 1903.

More broadly, it was altogether unrelated that, a few miles south of Kill Devil Hill, European settlement of America began in 1587 and that the first official call for American independence came in 1775 from Halifax, North Carolina, 120 miles northwest. Nor was it consequential that the Treaty of Paris, which confirmed our independence, was signed in April 1783 — a few weeks before the first manned flight — and in the same vicinity of the same city. It was not pertinent that North Carolina played a role in all of these apocalyptic events.

Or is it conceivable that notes from Roncommock's reed resonated most compellingly across the contiguous waters and woodlands where they were first heard? Is it conceivable that American independence, personal liberty, modern democracy, and the conquest of the sky emerged and evolved together because they complemented one another in releasing humankind from Old World parochialism, colonial tyranny, stubborn authoritarianism, and the pull of gravity? At some point along this sinuous pathway of history, does coincidence begin to sound curiously like destiny?

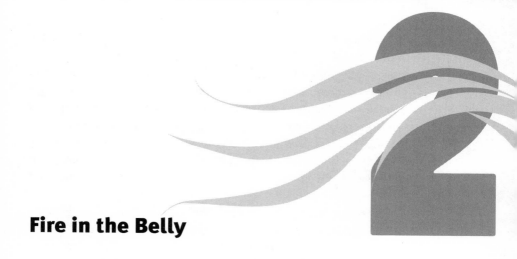

# Fire in the Belly

The first humans to fly did so in balloons, a mind-bending spectacle when first performed and still enchanting at the opening of the Third Millennium. Heavier-than-air flight, a far more complex challenge, required another 120 years. But the story of the early balloon and dirigible was an admirable achievement in its own right and one that played a role in the ultimate success of heavier-than-air flight.

Musing one day in 1782, Joseph Montgolfier of Annonay, in the Rhone River country of west-central France, pondered the futility of Spanish and French armies attempting to drive the British from Gibraltar. With its high, rocky terrain and powerful defenses, it was the strongest bastion on earth, as impregnable a fort as ever existed. It had no roof, but, of course, need fear no intrusion from the sky.[1]

As Montgolfier brooded, he idly watched smoke go up his chimney. Wouldn't it be grand, he thought, if smoke could be harnessed to carry things aloft? Well, he asked himself in his next breath, why not? With some old fabric, he made a small bag, lit a flame under it, and watched in fascination as the hot smoke sent it to the ceiling. Maybe his idea had merit.[2]

With his brother Etienne, Joseph in the next few months fashioned, from cloth and paper, an air bag thirty-five feet in diameter that carried its own flame under an opening at the bottom. (It is extraordinary how often teams of brothers made these advances.) They boldly announced that they would release their "balloon" in Annonay's main square on June 4, 1783. That day hundreds stood amazed as the flame heated the air in the object and drove it upward to a seemingly prodigious height.[3] About ten minutes later, its flame subsiding, it dropped gently onto a wall, where sparks from the dying flame burned the balloon. It was a short step for humankind but the first one en route to the moon and beyond; the science and art of aviation was born.[4]

The significance of the Montgolfiers' achievement was grasped almost immediately all across Europe. At the invitation of the French Academy of Science, Etienne hurried off on the 250-mile trip to Paris to give a demonstration. He arrived in early September to find that Parisian physicist J. A. C. Charles, having read of the brothers' experiment, had launched an unmanned balloon on August 27. Charles's hydrogen-gas contrivance rose about three thousand feet, traveled fifteen miles, and landed in a rural area, where frightened peasants managed to beat it "to death."[5]

Etienne, his performance scheduled at Versailles before Louis XVI, fashioned an immense and gorgeously decorated balloon forty-one feet in diameter, almost three times larger than that of Charles. He rigged it with a basket, in which he placed a sheep, a duck, and a rooster. Held on September 19, the exhibition lasted only a few minutes owing to a tear in the fabric. Spectators were thrilled and all passengers returned safely.[6]

Montgolfier rebuilt his balloon to a diameter of forty-five feet and a height of seventy-five, with a basket capable of carrying humans. On October 15 François Pilatre de Rozier, the designated driver, rode the tethered basket to an altitude of eighty feet before descending gently after about four and a half minutes aloft. A human had flown just 134 days after the first unmanned success, with American ambassador Benjamin Franklin among the admiring spectators.[7]

The Western world was electrified by this triumph of science over the forebodings of mystics. And well it might be; Paris had just witnessed a revolution whose consequences were perhaps more far-reaching than the mighty upheavals that erupted in those same streets on a July morning six years later.

Hot-air and hydrogen-gas ballooning soon enveloped Europe and America. Manned flights were made in England, Spain, and Austria in 1784, in Germany the following year, and by French aeronaut Jean-Pierre Blanchard in Philadelphia on January 9, 1793, with President George Washington on hand. In October 1797 Paris watched a successful parachute jump by André-Jacques Garnerin, a feat that would make ballooning more entertaining and possibly less dangerous. Meanwhile, Blanchard and others experimented with ways to guide flights, a challenge so daunting that it would outlast the eighteenth and nineteenth centuries.[8]

During the long wait for directional control, balloons found few practical uses. Tethered, they were employed for military surveillance in some nineteenth-century wars, including the American Civil War. There were competitions for the longest and highest flights, but these were chiefly circus attractions. Not long after the first Paris exhibitions, aeronauts were spellbinding crowds by ascending on horseback or astraddle a pair of alligators, or dropping small animals from great heights by parachute.[9]

As late as 1908, the powered dirigible (an elongated, steerable balloon) was considered the more promising vehicle for useful flight. A few Kitty Hawkers probably saw their first manned flight at an Elizabeth City balloon exhibition at the Albemarle Fair in 1903, two months before the Wright brothers first flew. And the balloon went much higher and farther.[10]

Some North Carolinians undoubtedly were among the crowds that saw the first balloon flights in the South, at Norfolk, Virginia, by impresario Phineas Parker in 1794, only a year after Blanchard's performance in Philadelphia. Passengers paid to ascend with Parker and marvel at splendid Chesapeake Bay vistas; they were led to believe that ballooning fostered good health by wafting them far above the noxious effluvia of city streets and malignant swamp vapors. Joseph G. Rea of Murfreesboro discovered a reason why he should have bought a ticket when his pocket was picked while he was ogling a New York exhibition in 1825.[11]

The date of the first manned balloon flight in North Carolina is unknown, but it probably occurred around midcentury. As late as the 1840s, an entrepreneur might gather a rural crowd by inflating and releasing small balloons. Even in Raleigh, citizens in 1847 settled for the release of a large unmanned balloon from Union Square to celebrate a visit by President James K. Polk to his native state. Chapel Hill students hailed Zachary Taylor's election to the presidency in 1848 by sending up, "with much grace and beauty," a homemade paper balloon thirteen feet tall and eighteen feet in circumference. Those who had never before seen a balloon thought this spectacle altogether miraculous.[12]

By the mid-1850s balloon exhibits were in demand across the state. Two flights in Raleigh by "Professor" George Elliot in late 1854 were said to be the 111th and 112th of his twenty-year career. One of his Raleigh ascensions left a permanent impression on Elliot's life. As he floated "near a cluster of cabins occupied by colored people," an old woman, "with outstretched arms," cried "'Come, Master, come Lord, I'se ready.'" The incident so touched Elliot that it made "a complete revolution," he later admitted, "in his [skeptical] religious sentiments." His next flight, in his "aerial car, Isabella," was for a Moravian crowd at Salem in February 1855; he descended at nearby Waughtown.[13]

Some Tar Heels were initially appalled at the idea of men and women ascending unbidden into the realms of the Almighty and were loath to tolerate such exhibitions in their communities. When Everett's show appeared in Statesville in 1859, its proprietor bought space in the *Iredell Express* to offer reassurance that "persons of the strictest religious principals [*sic*]" should "feel no repugnance in witnessing . . . the performance."[14]

It was a long time before North Carolinians became familiar enough with balloons to want to ride in them. The earliest on record is, perhaps, Vernon Henry Vaughan, senior yearbook editor at the University of North Carolina in 1859. Vaughan, later Utah's governor, read of the plan of Thaddeus S. C. Lowe, a premier balloonist, to cross the Atlantic from New York. He wrote to urge Lowe to take him along, but the trip was canceled because of technical problems. In 1861, however, Lowe flew over western North Carolina in an odyssey from Cincinnati to Charleston, South Carolina.[15]

Among the few practical uses found for balloons was the depiction of aerial panoramas by photographs, watercolors, and oils. In 1872 artist C. N. Drie rode in a tethered balloon to draft a multicolored "Bird's-Eye View of Raleigh" from a perch east of the present Pullen Park area. Some balloonists collected data for mapping, surveying, and weather conditions.[16]

In 1881 a couple at Cove Creek were reported to have wed in a balloon. The nuptial basket "became unmanageable before the bridal party came to earth, so that they barely escaped with their lives, passing the bridal night tossing about . . . , the bride being deathly seasick." (Another member of this interesting family was said to have been married first in a diving bell, a second time by telegraph, and a third as part of an amateur theatrical.)[17] The clan's name was not identified.

Tar Heels, like others across the nation, in time grew jaded with mere ascensions and demanded greater thrills. Performers answered with delayed parachute jumps, acrobatics on bars hanging from the baskets, night ascensions with fireworks and lamps, and other daring and innovative feats. Balloonists, now major attractions at carnivals and fairs, were admired and envied by spectators of all ages and circumstances. But profits were slim, and donations were usually solicited by passing a hat before an exhibition. In a typical case, a parachutist at Wilson in April 1897 made $4.80.[18]

Among the best shows were those at the North Carolina State Fair, including a memorable display staged by a Mr. Ward in 1890. As the *Raleigh News and Observer* described it, "the great bellying canvass began to rise, and then a tremendous cheer arose from the thousands of throats as it mounted skyward." The balloon seemed "like a great pear cut off from the stem. . . . As she rose she drew up after her a rope, to which was attached . . . the canvass parachutte [*sic*] . . . similar to a huge closed umbrella shaft; and some thirty feet below it . . . the bar which served as a support for the daring aeronaut."[19]

As Ward headed westward, the crowd sensed danger "and breath

Artist C. N. Drie drew this "Bird's-Eye View" of Raleigh from a balloon in 1872.
North Carolina Division of Archives and History, Raleigh.

was suspended. . . . Every eye was rivetted upon him, and every heart stood still in anxious solicitude." At about seven hundred feet, Ward detached his parachute from the rope and dropped hundreds of feet before it gradually began to open. Then, with "some suddenness it fully opened like a great umbrella and the rapidity of the descent was arrested. It is enough to say that every spectator breathed more freely" as the danger passed, "and the beauty of the performance was realized." Relieved of its burden, the balloon itself put on a show, "jerking in successive collapses great columns of black gas, until, fully emptied, it fell with a rush to the ground," Ward landing a hundred yards from where he had taken off.[20] Though limited by the inability to steer, a daredevil balloonist could captivate the audience and earn a decent living.

Interest increased when a woman was in the basket. In 1894 Miss Ruby De Veau and "Daisy" (a Skye terrier, of course) made jumps from her balloon at New Bern's fairgrounds. She came down, unharmed, in a treetop, the pooch negotiating the descent more successfully. By now it was routine for balloonists to carry advertising banners for local businesses and, at night, hoist fireworks displays, illuminate their balloons, and drop red flares to guide landings. They wore life preservers in case they pitched into water.[21]

With risk came danger. When high wind kept a balloonist at Salisbury from "hanging on a stick at the bottom of his balloon, by his hands," he had to make up for his failure. On his next effort, at Concord, the showman hung suicidally "by his toes to a cross bar" half a mile high. A sudden deflation at low altitude sent him plummeting into Si Harris's lot on Depot Street—shaken but not badly hurt. "It was a good show," said the *Concord Sun*, "and everybody was excited and pleased."[22]

With danger came injury and death. C. K. Perry's balloon collapsed at a Mount Holly show in 1889 before he could deploy his parachute. The balloon, dropping slowly at first, picked up speed and, in the last thirty feet, fell like a rock. Hitting the ground on his back, Perry was found unconscious and bleeding from his nostrils, ears, and mouth. Whiskey and a hypodermic needle roused him, but he died a week later in Charlotte, probably the state's first aviation fatality. Even spectators were not immune from injury. Two years

later a Raleigh African American, Anthony Jordan, caught his foot in
an ascending balloon's rope and fell fifty feet. He survived despite serious injuries.[23]

American contributions to ballooning included the invention by
the country's foremost balloonist, John Wise (1808–79), of the air-
release panel in the balloon's fabric, giving the aeronaut more control
of deflation and allowing a safer, more accurate landing. Thaddeus
Lowe's aim in proposing to cross the Atlantic in 1859 was to inaugu-
rate transoceanic mail service, but his balloon split while being filled
with air, and the coming of the Civil War ended further trials.[24]

## BYE, BYE, BIRDIE

In North Carolina, ballooning made human flight familiar to the
public from the mountains to the sea. Though it was confined
mainly to those with the money for the necessary gear and the daring to use it, ordinary folk remained keenly interested bystanders
and, occasionally, enthralled passengers. But in 1883 ballooning was
given a starkly human face by a story that swept across the state and
nation and even abroad. It concerned two-year-old Birdie Elliott and
her astonishing balloon adventure.[25]

Birdie, as identified to the *Greensboro Patriot*, was the child of Mrs.
Robert Elliott, a Texan who brought her to the resort town of Morehead City for a vacation and, it was hoped, recovery from a fever. A
"bright and sweet little cherub," she soon became a favorite of the
Atlantic Hotel resort set. On the morning of June 28, she was near
the hotel with "Major Hawkins of Alabama" when an Italian street
vendor passed with a large cluster of balloons. The major took a cord
of them, tied it around Birdie's waist, and gave her a playful toss into
a strong wind. To his horror, she was soon floating over housetops
toward Bogue Sound and the nearby ocean. At the major's shout of
"Great God! she is gone!" sportsmen raced to their boats.[26]

Some twenty men gave chase as, "with outstretched hands the
little angel could be heard distinctly calling 'mamma! mamma!
mamma!' until her voice became drowned by the whistling wind."
She was soon six miles east, sweeping seaward at forty miles an hour.

As distance widened between the boats and the helpless child, Charles H. Voorhees, of Greensboro, who had a Smith and Wesson rifle in his boat, began the dangerous and ticklish task of trying to burst the balloons without injuring the child. His fifth shot hit one and several more shots punctured others, though no one could tell whether Birdie herself was hit. She began drifting downward as the remaining balloons took her toward Harker's Island, eight miles east of Morehead City. The boatmen watched anxiously as she fell to the sand, fearful that she might be shot or injured on landing.[27]

Rescuers found Birdie playing contentedly with seashells, which she held up and announced calmly: "'Dese sels for mamma.'" The rescuers rushed her back to the Beaufort wharf, where they found Mrs. Elliott in a faint. But the reviving mother was overjoyed to find Birdie safe and unharmed. The dockside rang with cheers of admiration and relief.[28]

Readers who did not associate the name "Birdie" with that of the celebrated balloonist Elliott were taken in; hundreds of people across the country sent wires of joy and congratulation. Several gun makers, one of them in England, presented Voorhees with new rifles in token of his lifesaving marksmanship.[29]

It was soon revealed that the entire story had been concocted by Dr. G. W. Blacknall, proprietor of the Atlantic Hotel, as a publicity stunt and scripted by Pulaski Cowper, Governor Thomas Bragg's former private secretary. By mentioning fictitious genteel guests from Texas, Alabama, and across North Carolina among the dramatis personae, Blacknall depicted his hotel as a widely known and popular spa. The tale was such a sensation that he had to abscond for a time, reportedly to California. Rumor had it that, on his return, he would be sent aloft with a large cluster of balloons tied to each wrist, which Voorhees vowed not to shoot. Newsmen ranked this as one of the best newspaper hoaxes of the era.[30]

Probably North Carolina's most accomplished native balloonist was Robert Henderson Roper. Born in the village of Roper in Washington County in 1862, he left home at about age ten to join a circus. During his thirty-five-year career Roper worked with the best-known shows of his time, including the Barnum and Bailey Circus and Buffalo Bill's Wild West Show. In the early years parachutes were unavailable in America, so he could descend safely only by dousing the fire. But Roper's chief claim to fame was that he was the first American to jump from a balloon in a parachute of his own making. He was with the Van Tassel Show at Memphis in the 1880s when the owner devised a way to make jumps more interesting, an old idea that, because of the danger, had died out in America. Under Van Tassell's direction, Roper constructed "an umbrella-like thing," first descending with it into the Mississippi River.[31]

Ballooning produced North Carolina's first female aeronaut in 1892. In May, Frank Zelno brought his balloon exhibition to Charlotte, delighting spectators with headlong free falls before opening his chute. A fifteen-year-old in the crowd, Delia Jaquin, was impressed enough to ask to join the troupe. Being "rather petite, a brunette, and quite pretty," she was hired, with the written permission of her parents. The agreement was that she would make no ascension "for some time," and then only with Zelno.[32]

That was on Wednesday. On Saturday at Winston-Salem, Delia, now styled "Little Dot Zelno," soloed and released an imported Russian guinea before parachuting into a treetop from four thousand feet. A reception and bouquets from ladies and gentlemen helped fix Delia's mind on a career in ballooning. Surviving a collapsed balloon at five hundred feet over Baltimore, besides other mishaps, she visited her home in July, vowing that she would not give up her "career as an aeronaut for anything." She had now made fourteen jumps without serious injury. At Wrightsville in August, Little Dot performed a "double parachute leap," which meant cutting loose from a smaller chute that carried her little dog "Gyp." Zelno predicted that Dot would "become famous as an aeronaut."[33]

During the show's stay in Wilmington, manager R. C. Hutchison bought two bales of local sea island sheeting to make "the largest hot air balloon in America." His 90-foot-tall creation, requiring 1,400 yards of locally made cloth, was completed in a few days. It was christened the *John R. Nolan*, after the owner of a Wilmington railroad line. On August 19 it rose 3,000 feet, where Dot and Gyp jumped, then on to 8,000, where "Mademoiselle Leona" jumped. Dot landed, unhurt, 30 feet up in a live oak and was rescued with a ladder.[34]

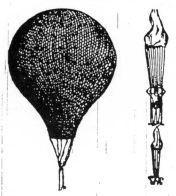

# Grand Balloon Ascension

## AT WRIGHTSVILLE

# TO-DAY!

FRIDAY, AUGUST 26th.

THE GRACE SHANNON BALLOON COMPANY
NO. 2 WILL CLOSE THEIR VERY SUCCESS-
FUL ENGAGEMENT AT WRIGHTSVILLE
TO-DAY UNDER THE MANAGEMENT OF
PROF. R. C. HUTCHISON, AERONAUTIC
ENGINEER.

## LITTLE DOT AND GYP

WILL MAKE THEIR FAREWELL ASCENSION
AND PARACHUTE LEAP AT 6 P. M. TO-
DAY, AND IT SHOULD BE WITNESSED BY
EVERYBODY, AS IT WILL BE MADE THE
CROWNING SUCCESS OF THE ENGAGE-
MENT.

THE FARE FOR THE ROUND TRIP OVER
THE WILMINGTON SEACOAST RAILROAD
WILL BE ONLY 25 CENTS.        aug 26 1t

"Little Dot" of Charlotte and dog "Gyp" perform in "America's largest balloon" at Wrightsville, August 26, 1892. From *Wilmington Messenger*, August 26, 1892.

The most celebrated North Carolina balloonist was James W. Smith, a farmer of Cypress Creek township in Bladen County. While he was watching a balloon being inflated at the Cumberland County Fair in Fayetteville on October 12, 1912, a gust of wind threatened to blow it away. Smith joined the effort to hold it down by its ropes. His ankle became entangled in a rope, and the balloon, suddenly ascending, jerked him up feetfirst "like a rifle ball." Hanging precariously by a foot, the fifty-two-year-old Smith reached for his knife to cut himself loose, but it had fallen from his pocket. He caught a guide rope swinging by and was able to right himself at about 1,500 feet. Meanwhile, some onlookers commandeered an automobile and sped after him as he vanished beyond tall pines to the east. Two miles up the road, they met him returning on a bicycle. He had pulled loose from the balloon as it touched down in a pasture before sailing off into space. He had borrowed the bike from a farmer and seemed no worse for the incident.[35]

Smith's story and name flashed along national telegraph wires, and he found himself a celebrity. Within a few days, however, he suffered an acute nervous reaction to his harrowing ride, with nightmares and at times even unable to speak. When the condition persisted, making it impossible for him to work his land, he hired a lawyer and sued the Cumberland County Agricultural Society, which sponsored the fair. The state supreme court found in his favor, but the case was settled out of court for an undisclosed sum. Smith lived to a goodly old age before his death in 1932.[36]

## JAKE HILL'S CIGAR

Balloonists, almost from the invention of the art, tried to guide themselves in flight. In 1784 Jean-Pierre Blanchard tethered a boat-like basket under his balloon. It had a rudder and big double paddles that opened and closed automatically as they were worked. Others tried elongated balloons with propellers, paddle wheels, and sails or obtusely hitched large birds to the balloons for propulsion—all without success.[37]

On September 24, 1852, a brilliant young engineer, Henri Giffard, achieved the first dirigible flight, near Paris. Slung under his enormous 144-foot-long balloon was a platform carrying a 330-pound, 3-horsepower steam engine with a 3-bladed propeller. A rudder was attached to the basket. Avoiding any attempt to fly into a brisk wind, Giffard made wide circles up to 6,000 feet high. He was the first balloonist to employ a motor in flight and to maneuver his ship with it. A premature death ended his promising research.[38]

Despite Giffard's success, dirigible flight was still a half century away, though airships sometimes achieved short flights during the next several decades. On August 9, 1884, *La France*, with an eight-horsepower electric motor, flew four and one-quarter miles at Paris, making the first circuit back to its point of departure. Further progress in fabrics and lighter, more powerful engines came in the final years of the century. But a persistent problem—lateral stability, or controlling side-to-side tilting—needed to be solved before dirigibles would have practical value.[39]

At the turn of the century North Carolina had at least two inven-

**Sketch of Carolina Central Fair in Greensboro, 1902. Note the obligatory balloon. From *Greensboro Daily News*, n.d.**

tors of its own who competed in the race to master stability and build the first successful airship. The *Norfolk Virginian-Pilot* claimed in late 1903 that both of them, tenant farmers, had flown before Wilbur and Orville Wright. The farmer-aeronaut, Jacob Aaron "Jake" Hill, lived in the rural Black Mountain vicinity of Stokes County. He had followed with avid interest the work of Europeans and believed that he could improve on their performance. Born in Davie County in 1862 the son of a Confederate soldier, Hill moved north to Forsyth in 1880 with his younger brothers, William and Henry. In 1891 they rented a farm in adjacent Stokes County, near Vade Mecum Springs, a tiny resort community. A decade later, Hill's ideas on the construction of airships began to generate press comment.[40]

Like many prospective aeronauts, Hill was intrigued from boyhood by bird flight, especially that of buzzards. A friend recalled him "chasing around under buzzards in a pasture, studying the way they fly. . . . It was all he talked about." Large birds flap their wings scarcely at all to remain for long periods aloft on currents of air; they made human soaring seem, with the right equipment, alluringly simple.[41]

The first public notice of Hill's work appeared in Stokes County's *Danbury Reporter* in March 1901, when he was in Danbury scouting for investors and legal help to take out a patent. Hill, said the paper, "claims to be the inventor of a machine" to solve "the problem of aerial navigation." Though declining to describe his machine, he revealed that it embodied "three distinct principles which render it absolutely practical, and . . . it will fly to any height or distance . . . against any sort of upper air current, with perfect safety." He called it "the most valuable invention of the age," but, as a tenant of small means, he needed help to bring it to reality. If the federal government showed no interest, he would approach the English. In late 1901 Hill announced that he had "obtained pecuniary assistance from wealthy men in Winston-Salem" and "gone North" seeking more.[42]

Three months later, Hill was far enough along to declare that he would enter the flying machine contest at the 1904 Louisiana Purchase Centennial in St. Louis. The grand prize of $100,000 for the best airship would be his, especially since all his rivals were likely to be mere balloonists. But his trip north yielded no further support. Some dismissed Hill as a bucolic crank, but they underestimated him. Even the *Charlotte Observer*, aviation's foremost champion, at first guffawed that his machine had come along just in time to keep the paper from buying an "airplane" advertised by Connecticut's Gustave Whitehead. It would now "wait a spell on the Stokes county man."[43]

On March 9, 1902, a Winston-Salem paper, at the site of America's first fictional air travel, commented on Hill's visit there a week before. The inventor had arranged for a working model to be built by the Ninth Street firm of W. J. and H. J. Hege, whose business included ironwork, bicycle, and sporting goods departments. The Heges were a family of mechanics credited with advances in cotton presses, circular saws, brick making, and other elements of the building trade. Hill's six-foot-long model would be cigar-shaped with a car underneath and a rotary fan on each side of the undercarriage.

**Jacob Aaron Hill as a young man.**
**Courtesy of Reverend W. Stephen King.**

It would somehow be steered by balloons and a propeller. A fully tested balancing system ensured easy management.[44]

Hill's dirigible would be similar to that in which Alberto Santos-Dumont had thrilled Paris for the last three years, but Hill had "overcome the great drawback" of such machines, namely, how to control them in flight. His mysterious balloons, operated "by a most ingenious device, controlled the machine at all times." "'I have watched,'" he said, "'every move of Santos-Dumont'"; his trouble "'is the controlling apparatus, and I believe that I have scored a great success in that respect.'" It was "'the feature I have worked on the hardest.'"[45]

Plans were made to form a stock company and build "a large number of machines." By May 5, the Heges had completed the model and Hill had applied for a patent. He would use a gasoline motor rather than electric. The full-sized ship could operate as low as four feet above the ground or as high as the operator wished or dared go.[46]

Newspapers statewide began to pay attention to his machine and follow its progress with regular reports. A Winston-Salem editor proposed that, although "[t]he world may be looking toward Santos-Dumont for aerial navigation," it should "not lose sight of our own Mr. Hill in the race." Businessmen, especially, were invited to stop by Heges' and see the model. (The Heges were fulfilling their claim to "what you can't find elsewhere.") The *Winston-Salem Sentinel*, noting that Hill had good photographs of the model, inquired about taking a two-day trip from New York to Paris. Hill replied, "'I'll do it if you'll furnish me with a 90 mile wind.'"[47]

But the winds of fortune blew erratically. A demonstration of the model was announced for May 25, evidently at Heges'. Visitors to the week-long Salem Female Academy Centennial were invited to see it. A month passed before the press said anything more about the airship,[48] evidence enough that the trial, if held, was disappointing.

Hill would not be denied. By June 26 he had sold seven $100 shares of stock in his flying machine and was ready to start work on a full-sized craft. (At least one investor, T. F. Callaway of Stokes County, paid only $10 for his "one Hundred Dollar stock interest in [the] proposed flying machine," voidable in a year if the buyer were paid $100.)[49]

Stock certificate issued to T. F.
Callaway for Hill's flying machine,
1902. From *Winston-Salem Journal*,
October 20, 1968.

In June Hill moved to Winston-Salem to head "a company with sufficient capital to build a real airship." A "practical test" would come within three months. Since stock sales were meager, he offered to demonstrate his airship at the town that offered "the best inducements." Its ascension, he promised, would be "a great spectacular event." Income from the sale of twenty-five–cent tickets to "an immense crowd" would cover costs not met by the sale of stock. All the stockholders apparently were, like Hill, Stokes County farmers. The *Winston-Salem Union Republican*, apprised that these investors were as impecunious as Hill, reversed its favorable view and cast jocular scorn on his project. He was building "a great hovering, swooping, gyrating greyhound of the air," a "ring-tailed skyscraper, that you can get in, pull a bell cord and say 'All aboard for St. Louis, Danbury or Stomping Ground!'" In no time, "you'll be sailing about among the comets and tumbling stars."[50]

The airship was in trouble. Hill called a stockholders' meeting at Vade Mecum on June 28 and organized his company. No more was heard of either the company or its project for over eight months. The last public notice of the endeavor, in March 1903, followed a Reidsville editor's receipt of a letter from Hill in New York. Parties there were interested, "and they think it is a great thing."[51] But nothing tangible came of the trip.

Despite his trifling budget, Hill managed to build his dirigible. An old photograph in the possession of his heirs, the only physical evidence that the machine existed, shows that it was about thirty feet long. The gasbag was sharply pointed in its conelike forward section; a conical rear section, shorter than the front, flared into a fantail rudder. A stabilizing airfoil hung horizontally between undercarriage and air bag. The motor, which cannot be seen in the picture, may have been mounted in a way similar to the one for the latest Santos-Dumont ship, at the midsection of the undercarriage.[52]

Wires and cables looped around the ship, apparently to tilt its nose up or down. Its guidance system of small balloons and a propeller seems actually to have comprised four parasols, one at each corner of the undercarriage, ribs facing outward. The pilot sat behind a row of wheels and pulleys at the front of the basket. The parasols may have been meant to open and close to accommodate changing winds and air currents. Since no hot air is mentioned, it is likely that the dirigible contained hydrogen or some other gas. Whether it had internal balloons, or the body itself held the gas, is uncertain.[53]

Hill had defied the skeptics but Mother Nature was little impressed. With a larger gasbag, his dirigible might have lifted itself, but it is unlikely that he would have been able to control it. Family members recalled that it hovered at times in the backyard shed, where it was kept for years, but it never flew.[54] He worked with cheap materials and no technical help. If he applied for a patent, none was granted. The award to Santos-Dumont in 1902 of a $20,000 prize for flying from the Paris suburb of St. Cloud to the Eiffel Tower and back, as well as further successes by Santos-Dumont, Count Zeppelin in Germany, and others, may have diminished Hill's enthusiasm for his airship. But perhaps he found satisfaction in having come as close to the goal as his circumstances allowed.

For the rest of his life Hill worked for the Winston-Salem gasworks, sometimes employing his aptitude for inventing for his personal comfort. A novel reclining rocker that he built is a proud relic of his heirs. A 1904 patent enabled him to easily cut bite-sized pieces from his tobacco plugs. Hill died in Winston-Salem on November 17, 1927, living long enough to witness Charles A. Lindbergh's October visit to Winston-Salem in the *Spirit of St. Louis*. But he was gone when the *Graf Zeppelin* traveled across the Atlantic to the United States and back to Germany in 1928.[55]

## MEET ME IN ST. LOUIE, LOUIE

A remarkable article in the *Reidsville Review* on February 18, 1902, reports not one, but two airships slated for construction in the area, one, Jake Hill's, about forty miles west of Reidsville, the other in Reidsville itself. Louis Henry Kromm, a Baltimorean, went to Reidsville in 1900 to work for the Honduras Manufacturing Company, a maker of pads for horse collars. Establishing his credentials as a civic-minded businessman, Kromm, in April 1901, was initiated into the local chapter of the United American Mechanics and named to the committee established to acquire a new school building.[56]

Reidsville was a busy tobacco town of small mills and factories bustling with skilled mechanics and builders. When a claim was made in 1903 that the first bicycle built in the state was in Leaksville in 1869, Reidsville's J. H. Laster was quick to assert that his brother in

Hill's dirigible, ca. 1902. Courtesy of Reverend W. Stephen King.

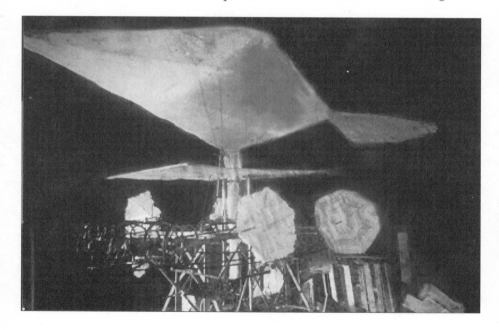

1868 had built an "ordinary," or high-front-wheel cycle, with iron-hooped wheels. Unlike Hill, Kromm had some technical background and access to neighbors with mechanical skills. In 1902, for instance, Reidsville wagon and carriage maker Fletcher Waynick built a water-cooled, five-horsepower gasoline motor in his blacksmith shop and attached it to a buggy. It was the second gasoline horseless carriage built in North Carolina—at a time when there were few anywhere. A carriage that operated without a beast of burden was little less astonishing than an airship might be.[57]

Kromm, the thirty-year-old son of a German immigrant, may have learned harness making at his father's Baltimore shop and possibly witnessed one or more of Samuel Langley's flying experiments on the Potomac River, beginning in 1896. At any rate, the *Reidsville Review*, one day after its account of the Hill and Kromm airships, carried a long interview with Kromm himself. "It is not," the editor declared, a balloon or an airplane, "but a substantially built air ship, fully equipped with the latest and most improved machinery." The model had been tested and it flew 595 feet, presumably with no motor, but perhaps a rubber band. The full-sized craft that he intended to build portended an immense forward leap in aeronautical science.[58]

Kromm's plan may have emanated from the fertile brain of Jules Verne. In his 1887 novel *Clipper Ship of the Air*, Verne featured a dirigible as big as an ocean steamer, with thirty-seven rotors for lift. It had a compartment for its engine, run by what might today be called fuel cells; four cabins (the pilot's was glass-domed); and a kitchen. Below the bridge deck was storage space for water, ammunition, and so on. It was said to fly at 125 miles an hour.[59]

The Kromm airship had a deck for passengers and another for machinery, steering gear, and the ship's management. A third deck, "above the machinery," held cargo. Built of aluminum, it was "lighted throughout," and all parts were made "to offer the least possible resistance in the air." It could be ironclad for use in war. Kromm called it "the only practical airship ever invented." Its power would come from a "powerful generator" of Kromm's invention, apparently more revolutionary than the airship itself. Its "perpetual power" would come "from the atmosphere" and cost "absolutely nothing." It was controlled just as pilots controlled ocean steamers and could fly "in any direction desired." Kromm reminded skeptics how, thirty years before, "when electricity was in its infancy, critics sought to disparage it; nevertheless it had shown its practicality by the great advances made since." His dirigible, he felt, deserved the same chance.[60] The proposed power generator sounded like a prevision, perhaps, of jet propulsion. Jules Verne would have loved it.

The inventor claimed strong financial support and seems not to have solicited local investors. He stated that he had organized the American (Air) Ship Construction Company in Baltimore but would build his dirigible at Reidsville's athletic park—presumably after the baseball season. The company's board was said to include Hon. Edwin A. Miller as president, Maurice Martin, Henry T. Kinsterman, Charles Gravel, and Kromm, a Baltimore team purported to have raised $100,000 in capital stock. New York investors were excited; a syndicate had offered him $40,000 for his interest.[61]

No explanation was given of how Kromm could form a wellfunded company and enlist the backing of prominent citizens in Baltimore while toiling over horse-collar pads in Reidsville. Nor was it clear why his project was unknown to the public until he had lived there for almost two years. No one seems to have asked if he had applied for a patent. Close inquiry, in fact, might have revealed enough hot air in his claims to launch a lead Zeppelin.

Kromm declared that his company was in touch with "the most scientific men in the United States and Europe" and had "letters and opinions" confirming that his machine would be a success. Indeed, it should be adopted by the federal and foreign governments to carry mail. President Edwin Miller, in Baltimore, was quoted as claiming that experts had examined the airship—presumably a model or blueprints—and believed that it promised to be the first practical dirigible. Kromm held that in moderate weather his machine could deliver mail, say from Reidsville to Charlotte, in under an hour, at a top speed of 120 miles an hour. The government would save thousands of dollars on mail between Washington and large cities. He hoped soon to see his ships "sailing from one city to another as regularly as the railroad trains leave city depots."[62]

The inventor revealed that, like Jake Hill, he would take his machine to the forthcoming St. Louis event, where his dirigible "will prove an easy winner. I know that it will pay us" to get it ready for the event.[63] The prospect of two North Carolina dirigibles at St. Louis, possibly the only ones there, was a grand notion for those citizens accustomed to viewing their state as "a vale of humility between two mountains of conceit."

Kromm told the *Reidsville Review* that he had become interested in flying after reading of Samuel Langley's glider flights. He had studied aeronautics for "a number of years" and perfected his model after a year of hard work. He offered to take the editor for a ride on his first trial, a proposal that the journalist rashly embraced.[64]

After the explosion of an airship in Paris killed its pilot in early 1902, Kromm announced that his machine was immune to such accidents because its power was nonexplosive. By mid-July he had been to Baltimore and back, purportedly having attended a meeting of his company's directors and won approval to take the airship to St. Louis. But this was the last press notice of Kromm's dirigible. Even his son Donald, contacted in 1992, had never heard of it. Donald Kromm recalled his father building a model monoplane in the 1920s but only as a recreation,[65] no doubt the closest he ever came to building an airship. His conception of his dirigible, however, was a magnificent one.

Louis Kromm soon returned to Baltimore, where he died on November 28, 1953, two days short of his eighty-second birthday.[66] He lived to see jets streaking across Maryland's skies and blimps guarding America's coastline, though mails still lumbered along between Charlotte and Reidsville on iron wheels. His intriguing scheme was a hoax from first to last. In a field where frauds were common, Kromm's project was evidently a singular aberration in a life of sober drudgery as a shop mechanic. What led him to pose as an aeronautical genius is anyone's guess; the idea was too fanciful for papers beyond Reidsville to comment on it.

With heavier-than-air flight still in the distance, the dirigible by 1903 appeared to be the means, more than automobiles or railroads, to revolutionize communication, transport, and warfare. A technological breakthrough in stability came on May 8 of that year, when Paul and Pierre Lebaudy, using a forty-horsepower Mercedes motor driving two propellers, flew a cross-country circuit, aviation's first, from Paris for twenty miles. A few weeks later they made a circuit of sixty-one miles. By this time, Count Zeppelin in Germany and others were poised to bring dirigibles into an age of commercial and military utility.[67]

The first dirigible exhibition in North Carolina was given by Lincoln Beachey at the state fair in Raleigh in 1908, a full two years before the state's first public airplane exhibit. Beachey's machine was owned by the Charles J. Strobel Company, a sponsor of shows across the country. The October 13 flight was pronounced an unqualified success, with the vehicle soaring "high . . . like some great bird of mythology." The next day Beachey flew over the city, circled the Park Hotel, and returned to the fairgrounds.[68] Dirigible technology was now securely beyond the talents of the Jake Hills and Louis Kromms. But could dirigibles satisfactorily fulfill Roncommock's decree? Many papers besides the *Charlotte Observer* did not think so.

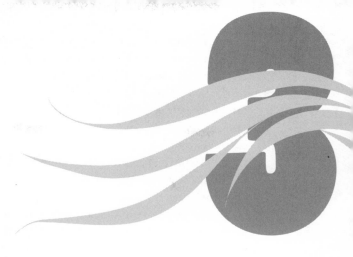

# Early Birds

Teasing "the secret from the womb of time" was the most intensely difficult and elusive pursuit ever attempted by humankind, one that consumed at least two millennia of study and toil. Thus no single boast of human prowess can stand comparison with the culminating success of manned flight.

Given the relative complexity involved, it is extraordinary that airplanes appeared so soon after the first horseless carriages and dirigibles. In the seventeenth and eighteenth centuries, Europeans, lacking a proper engine, sometimes tried heavier-than-air flight with their muscles alone. At the opening of the nineteenth century, however, the experiments and calculations of Sir George Cayley in England showed that, with materials then available, muscles alone could not power manned flight. He helped turn hopeful aeronauts to mechanical sources of power.[1]

Between 1809 and 1843 Cayley explored existing gas, steam, and internal combustion engines but found none that were suitable for aircraft. What was needed was a better ratio of weight to power, so he designed ingenious wings, horizontal and vertical rudders, propellers, and other features of modern airplanes. His work, generations ahead of its time, pointed the way to manned flight by heavier-than-air machines.[2] Thereafter, the stair steps toward the modern airplane were slowly but steadily scaled.

Cayley was still active when William S. Henson, also in England, worked out a complete description of an airplane, one that, if built, might conceivably have flown. His plan was for a fuselage to carry passengers, cargo, and a steam engine; large wood-framed, cloth-covered monoplane wings; and a tail and rudder attached under the frame. Twin pusher propellers were positioned behind the wings, not unlike those of the 1903 *Wright Flyer*. The design was patented in September 1842.[3]

The following year Henson joined manufacturer John Stringfellow to design and test a model, but without impressive results. Five years later Stringfellow built a model monoplane, 5½ feet long, with a 10-foot wingspan and a steam engine driving two 4-bladed propellers. He launched the model inside a large building by running it along a horizontal wire that automatically released it to fly. On the second trial, it rose until it hit a cloth suspended at the end of the building to catch it, flying most of the length of an 80-foot room. This was the first powered airplane flight. Later, in London, it flew 130 feet.[4]

Advances by the English prodded the French to take up where their neighbors had left off. In 1857 Felix du Temple built the first model airplane (weighing a pound and a half) to take off from the ground under its own power. His patent was for a plane of similar construction, with "a small boat" to carry the pilot, passengers, and a light steam engine; a twelve-bladed propeller would drive it forward. It had cloth-covered wings in wooden frames and twin rear rudders controlled by a wheel. Du Temple worked for twenty years to test his plane but never achieved flight—though contrary claims are sometimes made.[5]

In 1890 Clement Ader, of Gretz, France, is said on dubious authority to have flown his steam-driven airplane on an uncontrolled flight of 164 feet at a height of about 18 inches. His 650-pound batlike design had curved wings spanning 46 feet, a tractor propeller, and a 20-horsepower steam engine run by water and alcohol. This, he claimed, was "the first aerial apparatus piloted by a man to have flown . . . a short distance." His work seemed evidence enough that a successful airplane was not far off.[6] The French, as the world leaders in aeronautics, were widely expected to be the first to find the answers, though the remaining obstacles proved to be fiercely resistant to resolution.

Some inventors still thought that the last serious problem to solve was finding a suitable power source. American Hiram Maxim, working in England in the 1890s, built a 3½-ton airplane, 145 feet long, with 104-foot wings. This monster was meant to fly on power derived from two 180-horsepower steam engines, each driving a propeller 18 feet in diameter. The wonder is that, on July 31, 1894, Maxim sped his machine along a rail at forty-two miles an hour and actually felt himself jerked inches above the track before a propeller cracked and he cut power.[7]

Unlucky or misguided as they were, these men tended to be persons of skill and acumen, not cranks. No one knew how to build a machine that would fly, so one idea was about as good as another; insight came from trial and failure. Adequate motors were expected soon, but the question seldom addressed seriously enough was what to do once in the air—that is, how to control such a machine in flight.

Apart from Germany's Otto Lilienthal, who was killed in a crash in 1896, few had yet been able to fly even a glider, and there was little hope that his flimsy designs could be equipped with motors. Lilienthal had compiled data that ultimately led to the discovery of the proper camber, or front-to-back curvature, of glider wings. He made about two thousand glides in his career and was the first person photographed flying a heavier-than-air craft, an inspiration to the Wright brothers and experimenters around the world.[8]

Some Tar Heels may have first heard of heavier-than-air flight from an Alabama inventor who appeared among the regiments of the Army of Northern Virginia in the winter of 1864–65. R. O. Davidson, of Mobile, a recognized authority on aeronautics, envisioned a quick Rebel victory in the Civil War. As early as 1846, he had asked the Post Office Department for $2,000 to build a mail plane "for traversing the air like a bird . . . at the rate of 50 miles per hour." To perfect his Civil War version, he sought $20,000 and, by early March 1865, was said to have raised around $1,500.[9]

Davidson's warplane was supposed to fly by steam with "a man to guide it." It would drop shells on Union armies and "destroy or drive from our soil every hostile Yankee." He lectured units around Richmond and sought support from the beleaguered Confederacy for his "Artisavis" or Art(ificial) Bird. He felt that it would end wars, since no nation could withstand such a machine.[10] But his proposal came as the Confederacy collapsed in defeat and was never tested.

Despite the physical and social calamity of the war, North Carolina rallied in the 1870s to emerge as the South's leading industrial state. Machine technology was replacing handicraft, and railroads now carried valuable exports to pay for more advanced machinery. Progress in North Carolina was especially evident in tobacco manufacture, textiles, and the building trades. An exposition in Raleigh in 1884 enthralled visitors with a display of the latest machinery and processes, inspiring Tar Heels to redouble their efforts along these promising lines. The North Carolina College of Agriculture and Mechanic Arts (now North Carolina State University) was chartered in 1887 and was soon producing homegrown engineers.[11]

The *Sanford Central Express* lamented in 1889 that North Carolinians were "not a thoughtful, investigating, inquisitive people," as witnessed in the small number of patents granted to them.[12] Those with an inventive bent employed their talents mostly to generate thick batches of patents for cultivators, cotton choppers, cottonseed planters, bridle bits, and the like. Some of these devices worked well, but they were rarely designed with a view to diversify a farm-based economy or to answer the needs of commercial transportation.

There were already signs, however, of incipient change. A Charlotte firm by 1879 was turning out three-ton, tri-wheeled, steam-driven "road machines," invented by Charlotte's W. J. F. Liddell. Others around the state built similar vehicles for heavy-duty, short-distance hauling, rock crushing, grain threshing, and so forth. In the 1870s one Allred reportedly built a small steam engine, rigged it to a wagon, and drove the state's first automobile eight miles from Manchester to Fayetteville, panicking some who saw it. W. R. Self of Newton in 1886 invented a motor for "any sort of vehicle," including omnibuses and streetcars, for travel "at any desired speed."[13] Such straws in the wind evinced a quickening of the state's mechanical impulses, stirred mainly by the pace of modern industrialization.

The new technologies carried North Carolina toward delayed admission into the Second Industrial Revolution. The rapid movement of large cargoes was essential to the new system, and a few visionaries thought that air transportation might be a partial solution. Crude, experimental heavier-than-air flying machines began to pop up here and there, a technology that might supplement, or, in time, replace steamships, steam wagons, and trains.

## ARGOSIES OF THE SKY

A handful of airplanes, all unsuccessful, appeared in America between 1873 and 1903. The honor of constructing the earliest full-sized American plane is sometimes bestowed on Reverend Melville M. Murrell, of Panther Springs, Tennessee, who built his in 1877. But a more legitimate claimant is James Henry Gatling, of Hertford County, North Carolina, whose machine preceded Merrill's by four years.[14]

The earliest Tar Heel known to have designed a flying machine was an anonymous citizen of New Hanover County who lived near Wilmington in 1871. In October a Wilmington newspaper observed that "navigation of the air by means of machinery has long attracted the attention of scientific men." The latest one lived "in close proximity to this city, and the inventor, who soon intends to introduce his plans to the public, believes . . . with absolute certainty his work will be a success." His plans had been "submitted to scientific men here and elsewhere, and have . . . awakened an interest which is a tribute the demonstrator may regard as valuable." He would "organize a company to give . . . reality to the theory."[15]

The optimistic tone of the report typified many such claims over the next three decades, before the mysteries of manned flight were more fully analyzed and appreciated. The Wilmingtonian may have had some kind of dirigible in mind, in which case failure was not a foregone conclusion. But anything else was almost certain to fail in the years before the internal combustion engine and a sense of how any machine could be managed in the air. In any case, nothing more was heard of the project.

Other Tar Heels who were investigating the secrets of heavier-than-air flight soon produced more than paper or theoretical projects. On March 18, 1872, the *Raleigh Daily News* carried a feature on the Gatling plantation in Hertford County, birthplace of the machine-gun inventor Richard J. Gatling. The writer, a neighbor, noted that Richard's older brother, James Henry, was "at work on a machine, that is destined at some *future* day to eclipse the famous gun and fly triumphant over time and space."[16]

James Henry Gatling, born in 1816, was, according to local tradition, beguiled by flight from boyhood. He was said to have been a prolific maker of toys, flutes, animal traps, bows and arrows, and so on. He jumped from barn lofts with wings of fodder and made kites and paper and wooden model airplanes "in the shape of various native birds." On one occasion he reportedly ruined an umbrella when it "flopped inside out" during a jump he made from a barn loft. Gatling was too old to take part in the Civil War, though as a slave owner he had much at stake in the outcome. But the loss of his slaves may have kindled his own inventive ability. In 1871, nine years after his brother built his first machine gun, James Henry patented a novel cotton stalk chopper and a means to preserve wood.[17]

Richard Gatling was living in Indianapolis and, though he sometimes visited the family farm, seems to have had no part in Henry's airplane project. But his antebellum marine screw propeller—four blades set at angles of forty-five degrees about the axis of rotation—was rejected for a patent because John Ericsson's had already been approved. Richard's steam tractor and gunboat designs also may have influenced Henry's ideas. Even the celebrated rotating-barrel machine gun bore mechanical affinities with Henry's airplane. Inevitably, the younger brother found in Richard's success and celebrity strong incentives to pursue his own creative ideas.[18]

Folklorist F. Roy Johnson pieced together the story of Gatling's airplane mainly from elderly local people and Gatling progeny in the 1960s and collated their comments and descriptions. Among several published sources buttressing his inquiry was a letter in the *Norfolk*

*Journal* of March 18, 1873. The writer, a resident of Murfreesboro, the nearest town to the Gatling farm, mentioned a recent visit there and seeing "a flying machine in embryo, with which when completed an enthusiastic friend expects to navigate the air like any of the feathered tribe."

Local records disclose that Henry Gatling was industrious: an expert maker of wine and apple brandy; an erstwhile merchant and sawmiller; a skilled farmer, hunter and fisherman, and thwarted

Gatling plantation, Hertford County, where James Henry Gatling attempted flight in 1873. The loghouse in the foreground was the birthplace of machine-gun inventor Richard Jordan Gatling. From *Potter's American Monthly*, May 1879, 334.

Portrait of James Henry Gatling
by Oliver Copeland, ca. 1854.
Courtesy of F. Roy Johnson.

town builder; and a combative defender of his honor and integrity. He was educated in a local classical school. In his sixties, the dapper bachelor seemed ten years younger, and, at five feet seven, looked trim in his five-inch black chin-whiskers and favorite half-beaver hat, a popular midcentury style.[19]

By all accounts, North Carolina's first airplane had a profile similar to those of soaring birds and modern planes. Gatling's observations of buzzards apparently led him to suppose that power, lightweight materials, and a dash of control was a recipe for flight. His plane's fuselage was of light poplar, its triangular monoplane wings of white-oak splits an eighth of an inch thick and held in place by a wire frame. Wires connected the tips of the hinged wings to a cockpit lever so that the pilot could move them up and down in flight as needed. A front elevator caused the nose to move up or down; a vertical rudder, operated by the same lever, was attached at the tail. Twin wooden propellers, worked by a handwheel in the cockpit, would draw air into casings in front of each wing and force it out under the wings for lift. It was eighteen feet long, with a fourteen-foot wingspan and tricycle landing gear. Some witnesses labeled the machine "the turkey buzzard," presumably in honor of its inspiration.[20]

Like other tinkerers of his day, Gatling had little idea of the power it would take to fly such a machine. Steam engines provided the best

**Sketch by artist Bill Ballard, from descriptions by witnesses, of Gatling and his 1873 airplane. Courtesy of F. Roy Johnson.**

propulsion available but needed a heavy burden of water and fuel for operation, even for short distances. So Gatling proposed to fly with muscle power alone, trusting that the demands on his strength would diminish once his plane was airborne. This, at any rate, seemed to be the lesson of soaring birds. At one point, he revealed to a friend that, if his machine proved successful, he would pay a visit to his brother Richard in Philadelphia.[21]

In amply attested legend, Gatling tried his only flight on a Sunday afternoon in 1873. His course was to be over some woods east of the family farm to a road a mile distant (now Highway 258), where he would turn and fly back. He took his seat in the cockpit chair and had six farmhands push the plane off a twelve-foot-high platform protruding from his gin mill. With Gatling furiously cranking his wheel and working his levers, the plane reportedly soared over a four-foot fence, veered left, and struck a wing against an elm tree. It spun around and crashed, smashing the left wing but leaving its operator with only minor injuries. The distance of the flight was said to have been an improbable one hundred feet, just short of Orville Wright's first flight of 1903. Despite its failure, the machine was the talk of the neighborhood for years afterward.[22]

The experience did not warrant repetition. Gatling stored his airplane and never tested it again, though he is said to have considered installing an electric motor. While feeding his hogs on the morning of September 2, 1879, he was shotgunned to death, for a minor snub, by a deranged neighbor, William H. Vann. His plane burned with its shed in 1905.[23] It may be technology's loss that the Gatlings did not, like the Montgolfiers and Wrights, pool their talents to produce, for example, a rapid-fire warplane.

About sixty-six years after the fact, Guy Picot, a former neighbor of Gatling's, affirmed in a volume of county history that Henry had tried to fly his plane from the gin mill platform. "But," he says, "instead of winging its way skyward, it rapidly descended to the ground," almost killing the pilot.[24] The sketch of the craft shown above, which synthesizes eyewitness accounts, is probably distorted by efforts to make it look more like a modern plane than it is likely to have been. The wings, even if impervious to the flow of air through them (we hear of no wing coverings or varnish), were too small for the weight that Gatling meant them to bear. The fuselage, propeller casings, wheels, and elevator are all clearly too heavy to be lifted by muscles alone. The drawing represents a general scheme of the machine but, on the whole, not one that should be tested from a gin mill platform.

In any event, "before Tennyson dreamed of Argosies sailing through the sky," wrote Picot, "Henry Gatling dreamed of the flying ships. . . . He had a vision," one given substance "by the Wrights at Kitty Hawk." It is the vision of such dreamers, more than their failures, that deserve our notice.[25]

Other Tar Heels were at work on flying machines and sometimes gained wider notoriety. The progress of steamships and railroads fostered increased efforts to master the skies. A worthier machine, in fact, was already under construction in western North Carolina before Henry Gatling's death.

In January 1881 the *Charlotte Observer* announced that "Dr. Asbury's flying machine is completed," except for "a few finishing touches, and . . . next week sometime, . . . will sail from a point in the country, near Charlotte, and land in Independence Square. . . . Keep a sharp lookout over head next week, and you may catch sight of it as it sails into town." Though Asbury was reticent, the *Observer* learned that the plane was "buoyant in itself; the propelling power is to be supplied by hot air." Further hints established that it was an airplane rather than a dirigible.[26]

Dr. Daniel Asbury, planter, geologist, physician, and inventor, was born in Taylorsville, Anson County, in 1816, the same year as Henry Gatling, and became the well-to-do owner of a goldmine. In 1858 he bought a farm on Tuckaseege Road, near present-day Douglas International Airport, outside Charlotte in Mecklenburg County. It had "a splendid orchard and one of the largest vineyards in this part of the state." During the Civil War, Asbury and an associate, Dr. E. Nye Hutchison, invented a type of artillery shell capable of a "double explosion" that hurled the shell great distances, like the stages of a rocket. But trials at Charleston, South Carolina, failed and the project was dropped.[27]

After the war, Asbury developed a method of flue-curing tobacco that was widely used in North Carolina and Virginia into the twentieth century. In 1874 he patented an apparatus for "bleaching, making extracts, etc.," and, three years later, a novel brick kiln. He was also said to have experimented with a "drying apparatus." Five years after that, at age sixty-three, he tested new burning methods to reduce the amount of firewood needed and produce "even temperatures throughout the kiln." These projects indicate that he was strongly drawn to thermodynamics and its practical applications.[28] It was this interest that led to his airplane concept.

When the *Observer* broke the news of Asbury's intended flight from his farm seven miles west of Charlotte to the city's Independence Square, it noted that he had for "many years" pursued the study of "bird flight, air currents, etc." Construction of the machine had occupied the preceding two years. Asbury, Hutchison, and a Mr.

Martin, a skilled Charlotte mechanic, intended to launch the plane from the roof of a tall barn on a neigboring farm, "Enderly," owned by Dr. Sydenham Benoni Alexander, grandson of former governor William A. Graham.[29]

For undisclosed reasons, the January 1881 trial was postponed, but in late March the plane was said to be near completion and would make its flight within a week. Asbury announced his flight path into Charlotte in advance lest someone "might take it for a strange bird" and shoot at it. Witnesses called the craft "a beautiful piece of mechanism, with strong probabilities . . . for success."[30]

By now, some details of Asbury's "ship of the air" were being passed on by newspapers. Its shape was "something between a boat and the body of a bird," with thirty-foot canvas-covered wings "in imitation of a bird's wings." Its "big, torpedo-like body," "almost identically the shape of a bird," was "made of sail cloth and covered with ropes attached to a small basket underneath." The location of its steam engine is not mentioned, though it was probably in the basket. Wings, "on either side of the body," extended "out of the boat like bird wings." (Nothing is said of where fuel was to be carried, but it would presumably be in the basket with the engine and pilot.) The tail was also "like [that of] a bird."[31]

The Asbury plane derived power from two side propellers presumably driven by steam. The pilot, seated in the basket, worked foot pedals "like a sewing machine," causing the wings to flap and the machine to ascend. (The Asburys were said to own one of the South's first sewing machines, a Howe.) At a desired altitude, the pilot stopped working the wings, which, "being extended on a level, the bird-machine, with the aid of propellers . . . , sails through the air at any distance." Because of its precise balance, the pilot could descend just by leaning a bit forward, giving "the beak the right incline downward." A daughter of Asbury recalled that the craft was "an exact prototype of the Bleriot plane of the present day," the World War I era, hence a monoplane.[32]

Asbury's invention received wide notice in the press, some of it scornful. A wag in the eastern hamlet of Whitakers wrote to a Raleigh paper that he hoped the plane would be a success in time to enable him to airmail the editor a one-hundred-pound watermelon.

The *Philadelphia Evening Bulletin* assumed that it was some kind of balloon. The editor was himself "ready to slip earthly cables for a heavenward flight, if we should feel that the whole merit of the thing is not in the going up, but that something was left for the coming down." But a critic who warned Asbury that nothing without feathers could fly had no response when the doctor mentioned bats.[33]

Following a March 30 update, nothing more was heard of the plane for several years. Asbury became ill and, on March 16, 1882, died of pneumonia complicated by pleurisy. He was mourned as "a man possessed of much information and a considerable amount of genius." His associate, Martin, was said to be at work on the "propelling mechanism" when Asbury died, but evidently soon abandoned it. The plane was later dismantled and the bleached sheeting of its wings "used for other purposes."[34]

Asbury's name reemerged from time to time for decades. In early 1904, following the Wrights' first flights at Kitty Hawk, the *Charlotte Observer* inquired around town about whether the Asbury plane had flown, thus giving Charlotte what was claimed for Kill Devil Hill. Some said it did, but the best-informed sources denied it. Others claimed that it was still in storage in a building in town but were re-

**Dr. Daniel Asbury of Mecklenburg County, who built a steam-driven airplane in 1880–81. National Air and Space Museum, Smithsonian Institution, Washington, D.C.**

butted by those who maintained that it had long since been dismantled. Dr. Hutchison, the only person who knew the answers, perversely refused comment.[35]

It is impossible to evaluate the capabilities of the Asbury plane,
but it was the work of a hardheaded, practical-minded, and successful inventor. With what were probably flat wings, it was unlikely to
have flown in any meaningful sense, but it might have been capable
of lifting itself off the ground and traveling short distances under
some degree of control. Unfortunately, posterity has an inadequate
description of the craft and the precise role of steam in its operation.

The Gatling and Asbury machines were, so far as is known, the
only full-sized airplanes built in North Carolina in the nineteenth
century. But a rebuttal comes from coastal Beaufort County, where it
is said that the just claims of a Pantego man have never been properly recognized. Several people have sought to have a monument or
highway marker erected in his honor or to persuade the navy to
name its Lighter-Than-Air facility at Weeksville for him. His name
was John Maynard Smith.[36]

Smith was born at rural Wilkinson Station, three miles north of
Pantego, on March 12, 1875. His mother died when he was a few
months old, and he was reared in a three-room backwoods home by
his father, Daniel Isaiah Smith, who worked on his five-acre farm
and, off-season, at sawmills. With a scant year or so of formal
schooling, the son became engrossed in the flight of hawks and other
large birds and began working on a toy with birdlike wings. In 1889,
at age fourteen, Smith built a model biplane and looked for a way to
make it fly on its own power. From an old timepiece, he assembled a
clockwork mechanism that caused the plane to rise and fly around a
room. Neighbors and others heard of the model and often dropped
by to see it perform.[37]

This is about as much of Smith's story as we can assume to be reliable, but it is by no means the end of the tale. In a newspaper interview in 1955, Smith confirmed the preceding details and added that
he had built, on a similar pattern, a full-sized, single-engine plane
around 1892 but was too poor to spend $400 for a motor. Confident
that it would fly if he had financial aid, he wrote to President Grover
Cleveland to ask for federal help but received no reply. Not long af-

terward, according to Smith, two men, named Curren and Houtzen, visited Pantego, heard of his plane, and asked to see it. They looked it over and "told me it wouldn't work because the propeller was at the front end," an objection presumably derived from the placement of boat propellers.[38]

In the 1970s a friend of Smith's, Reverend R. M. Gradeless of Belhaven, decided that Smith had been robbed of his rightful place in history—for inventing the airplane. Gradeless wrote numerous magazine and newspaper articles on how Smith had built his machine as a teenager, only to have it snatched away from him by conniving strangers. In the Gradeless version, two mysterious well-dressed men visited the humble Smith home in Wilkinson Station, asked to see the clockwork model and its plans, stayed overnight, and absconded with both before dawn.[39]

Perhaps influenced by Gradeless, Smith then claimed to be "the true inventor of the airplane." In his 1955 interview, he had revealed that friends told him that Curren and Houtzen were "sent down by the Wright brothers," apparently to steal his secrets, "but," Smith had added, "I do not say so." Gradeless asserted that President Cleveland was shown a model of Smith's airplane and laughed because of its front-mounted propellers. He added that Smith almost perfected a perpetual-motion machine before that, too, was stolen. A friend of Smith's stated in 1994 that the motorless full-scale model plane was eventually given to "some man down Hatteras way" and had not been heard of again. Smith died at Belhaven in 1963.[40]

## THE COMING OF DR. CHRISTMAS

William Wallace Whitney Christmas, a name no longer recalled in his native state, was one of the most remarkable men in aviation history. Born in Warrenton, on September 1, 1865, he was the son of a reputable family. His father was proprietor of a distillery that turned out the noted "Old Christmas" whiskey at the rate of ninety-nine barrels a day. The son's great-uncle, William Christmas, was a state senator who surveyed and laid off the city of Raleigh, among others.[41]

Young Christmas, by his own accounts, built kites and studied bird flight in the 1870s, when Henry Gatling tinkered with his mono-plane down the road in Hertford County. A former neighbor of the Gatlings, Dr. L. Julian Picot, was by 1875 a Warrenton resident and well known to the Christmas family.[42] It may have been Picot who introduced the lad to the notion of artificial flight.

William was the oldest of three children of Confederate veteran James Yancey Christmas and his wife Rhoda Gaines. Soon after her death in 1879, the family moved to Washington, D.C. William later said that he attended St. John's Military Academy in Alexandria, Virginia, and earned bachelor's and master's degrees at George Washington and Johns Hopkins Universities, as well as a medical degree at Hopkins. As a Washington physician, he is said to have specialized in nervous disorders.[43]

By 1889, a year after the death of his father, the twenty-three-year-old had matured into a gifted and promising young man. A Washington journalist called Christmas "a born genius, . . . handsome and clever," who played guitar better "than any amateur in this city, and is self-taught." He had built a "trim yacht," the *Rhoda*, and was "a daring, fearless and expert sailor." Moreover, he supported his brother and sister on his salary as "a messenger in the treasury." His "chivalrous attentions" to his late mother's aged French maid reminded the reporter of "the days of romance and knighthood more than the bustle, rush and ingratitude of the present century."[44]

There was more. Christmas was a talented watercolorist and painter. His depiction of the *Monitor-Merrimack* ironclad battle, based on his father's eyewitness description, "won an exhibition" in 1889. Listed in the 1901 *Who's Who in American Art*, his work included Washington-area landscapes, one of which, *Deep Woods*, took second place in the Pennsylvania Academy of Fine Arts Exposition in 1911.[45]

By 1881 his interest in flight had become an obsession. He conducted experiments with huge kites to determine how much weight they could lift. His largest, eleven feet long, needed three men to hold it in a strong wind. On one occasion, the cord of a big kite caught his arm, jerking him skyward and cutting a gash in his arm. He was saved from serious injury by an alert companion, who grabbed him and pulled him down.[46]

Christmas learned "the principles of flight," he said, "from the . . .

experts. . . . I got corns on my back from lying on the ground studying birds." By baiting buzzards, he could watch them closely in
flight. He noted, for example, that the flexible wings of birds coun-
teract strain and relieve them from sudden shocks, suggesting that
flying machine wings should be flexible rather than fixed. The
"sculling" of flight feathers, he discerned, helps drive birds forward,
while tilting wings at high angles creates an upward force that slows
birds for landing. Using his training in physiology, he showed that
the amount of food eaten by birds provides too little energy for sustained flight, proof that currents of air furnish both support and
much of the propulsion for flying.[47]

During his heavier-than-air studies in 1898, he is said to have conducted "a lengthy correspondence" with Gustave Eiffel in Paris,
builder of the Eiffel Tower and an early expert on aeronautics. An aviation historian says that Christmas "made kites, model aeroplanes,"
and a glider to test his ideas and designs. In 1905 he reportedly gave
a public exhibition by flying the glider "from the base of the Washington Monument." He would soon build his first airplane.[48]

But little of this information on flight studies and experiments
has been verified from independent sources; indeed, historians have
grown skeptical of nearly everything said by and about Dr. Christmas.[49] (Perhaps more than is fully justified.) Since he will turn up
again in these pages, it is worth examining him closely beforehand.
That he was a genius of sorts is certain; that he was, in many respects, aviation's foremost Münchhausen is equally so. But the line
between the two is obscure.

A circumstance in his childhood may reveal something useful
about the man he became. He was a grandson of Myra Clark Whitney Gaines, a Creole beauty, who in the 1830s, owing to a complex
tale of paternal inheritance and deceitful lawyers, asserted a claim to
ownership of about a third of the land area of New Orleans. After
epic legal battles of over half a century, and six years after her death
in 1885, the case was decided in the U.S. Supreme Court in her favor,
and her estate was awarded about $600,000 in settlements. (She also
won many claims against individual property owners in the city, so
her compensation, mostly consumed by the suits, was far larger.)[50]

In the meantime, her family became irreconcilably divided over

**Dr. William W. Christmas, a Warrenton native whose aeronautical career stretched from the late nineteenth century until his death in 1960 at age ninety-six. National Air and Space Museum, Smithsonian Institution, Washington, D.C. (Videodisc no. A48316).**

the anticipated inheritance. In 1880 Mrs. Gaines set up her son, William G. Whitney, and son-in-law, James Y. Christmas, in a hair mattress business in Washington. But the enterprise, damaged by the partners' mutual ill will, soon collapsed. Whitney felt that "the attention Mrs. Gaines showed to the Christmas children" threatened his own interest in the prospective estate. The two men became so estranged that they would neither speak to one another nor eat at the same table. On June 25, 1881, James Y. Christmas fatally shot Whitney in a boardinghouse fracas.[51]

A contemporary account describes the pitiable condition of six children "standing around an open grave, three orphaned by the hand of the father of the other three, and united by the closest of blood." All "must look to Him alone, who can straighten out the tangled webs an adverse and inscrutable Fate seems to have woven around them." James Y. Christmas was accompanied to jail by his children, who "left with tears in their eyes." The court, however, found him to have fired in self-defense at an unbalanced assailant.[52]

William Christmas, who was fifteen at the time, had to adjust to the death of an uncle at the hands of his father, the loss of his mother and natal home, and responsibility, after his father's death in 1888, for his younger siblings. Nevertheless, the settlement of the Gaines estate made him a corecipient of $300,000, his third allowing him to pursue bird flight or cockleshells wherever they might lead.[53] But the trauma of successive tragedies at an impressionable age may have taken its toll in emotional stress. One may imagine that his genius developed within a perspective that rationalized contempt for other people and programmatic duplicity toward them. At any rate, his early history should be borne in mind as we encounter more fully documented phases of his extraordinary aeronautical career in subsequent chapters of this book.

Were there other, unrecorded, airplane experiments by North Carolinians before the Wright brothers? Contemporary written comment on Henry Gatling's machine seems to have been limited to two sentences in widely separated newspapers. Only because his brother Richard's fame made his birthplace something of a shrine was there any public interest in what went on there. John M. Smith's experiments were not recorded in the contemporary press. It seems safe to suppose that other Tar Heels built or at least designed flying machines before the close of the nineteenth century.

With the exception of Smith, the North Carolina experimenters were all patenting inventors. (Smith himself was an inventor but was never awarded a patent.) They also had in common a passion for studying bird flight and trying to duplicate it mechanically, a characteristic they shared with the Wright brothers. And, having progressed through the stages of muscle, steam, and clockwork, most of them were ready to turn to gasoline engines.

Neither Henry Gatling, nor Daniel Asbury, nor John M. Smith contributed anything to the evolving technology of aviation, but this was true of the great majority of experimenters everywhere. The work of the North Carolina inventors, however, reflects the grassroots fascination with aviation in nearly every state in the Union. Given this passion, it was likely that someone, before long, would get it right, perhaps even in North Carolina. The Conjurer's decree, in fact, was about to be fulfilled.

# As the Surfmen Tell It

## THE SORCERER'S APPRENTICE

William J. "Bill" Tate, acting postmaster and assistant weatherman at Kitty Hawk, was an up-and-coming young man but with no inkling that he was soon to become a historic figure. On August 7, 1900, while sifting through the mail at the post office in his home, he noticed an envelope, with a Dayton, Ohio, postmark, directed to the weather station. He opened it and read that a Mr. Wilbur Wright wished to know "whether he could find in that section level plains [free of] trees . . . to make some experiments in . . . kite flying."[1]

Tate showed the letter to chief weatherman Joseph J. Dosher, who dashed off a brief reply recommending that Wright bring a tent, since rooms were scarce at Kitty Hawk. He also advised Tate to respond more fully to the inquiry. Bill Tate had recently read an article on flying and knew that kiting was a well-worn path toward the goal of powered flying machines. In any case, Wright's proposal might liven a season promising no more than a shipwreck or two to break the all-encompassing monotony of the Outer Banks. His letter was warm and inviting; it promised everything Wright could want and betrayed his eagerness to entice the Ohioan to Kitty Hawk.[2]

"You will find here," he wrote, ". . . a strech [sic] of sandy beach 1 mile [wide] by five [long] with a bare hill [Lookout Hill] in the center 80 feet high not a tree or bush any where. . . . [O]ur winds are always steady generally from 10 to 20 miles" an hour. Kitty Hawk had thrice-weekly mail from the mainland, a telegraph line, and board with private families. "I will take pleasure," he added, "in doing all I can for your convenience & success & pleasure, & assure you of a hospitable people when you come among us." But the best weather was from mid-September to mid-October, so Wright should, if possible, come at once.[3]

The letter that drew Wilbur Wright to Kitty Hawk carefully omitted such details as the three months left in the hurricane season, the fierce northeasters, and wind-driven lateral rainstorms, as well as mosquitoes, sand flies, gnats, and other critters that held unchallenged sway over the beach. He probably knew that he had oversold his lonely sand spit and wrote that R. G. Allen of Dayton's weather bureau had formerly been stationed there and could be helpful.[4]

Wright arrived unannounced at Kitty Hawk on September 13, 1900, without even replying to Tate's letter. After passing through Norfolk by train, he had spent the night of the tenth in Elizabeth City at M. J. Sawyer's Arlington House Hotel. The next day he hired the only available vessel, Israel Perry's rotting shantyboat, the flat-bottomed sloop *Curlicue*, for the forty-eight-mile trip to Kitty Hawk. His brother, Orville, would join him after closing their Dayton bicycle shop for the season.[5]

Elijah W. Baum, a lad of fourteen, was sailing a model boat by the bay on the morning of the thirteenth and recalled that a neatly dressed young man stepped off the boat at Kitty Hawk wharf and asked where Tate lived. Baum led him to Tate's home "a mile up the sand road." It was a five-room, two-story, frame house and, like most others in this sand-whipped shore, void of paint.[6]

"In walked a bare-headed man," Tate recalled, "with a fringe of hair skirting his ears and the back of his head. 'Is this Mr. Tate? . . . I'm Wilbur Wright from Dayton, Ohio.' 'How the devil did you get here?'" asked Tate. "'Oh, heavens don't ask me,'" Wilbur replied, "'I came down with Israel Perry.'" He "proceeded to unfold a tale of hardship" about his voyage, "how the miserable boat ran for harbor in a blow, and how he could not eat the provender" on it and had had no nourishment for forty-eight hours. "His . . . description of the rolling of the boat and his story that the muscles of his arms ached from holding on . . . called for a remedy at once."[7]

"It being after our breakfast hour," Tate continued, "about 9:30 A.M., a hasty fire was made in the kitchen stove." His wife Addie set out a hot breakfast of eggs, ham, and coffee, and Wilbur "did his duty by them." The next day, when Wright expressed concern about contracting typhoid from the Tates' surface-well water, Addie began boiling a daily gallon and leaving it in a pitcher in his room "to safeguard him."[8]

Wright's baggage and crated glider kite soon arrived from Elizabeth City on Captain Franklin H. Midgett's freight boat, the *Lou Willis*. Bill Tate offered him room and board, so they "struck a bargain" for four dollars a week. Addie, recalls a daughter, "told Papa that our fare would be too plain for city gentlemen." But Wright overheard her and ingratiatingly announced: "Mrs. Tate, we shall be guests in your house; your fare will be our fare."[9]

News went around that Wright "was inventing a flying machine." The hosts and their neighbors looked on in awe and amusement in the following days as Wilbur Wright engrossed himself in woman's work. Pitching a canvas lean-to in the front yard, he "unfolded bolts of cloth on the . . . porch, cut them into suitable lengths, and asked to use the family sewing machine." Addie lent him her treadle Ken-

William J. "Bill" Tate of Kitty Hawk, host and aid to the Wright brothers from 1900 to 1903. Courtesy of Daniel Grady Tate.

more to sew together segments of wing cloth and did some herself. Locals long recalled Wright "on his knees in the yard at work," with his materials on the nearby porch. Orville arrived in late September with a bigger tent to find Wilbur, plagued by hot weather, still at work on the glider's elevator and lower wing.[10] The villagers were enjoying a fine show.

Bill Tate became Wilbur's first North Carolina convert. Seated with him a few days later "on the steps of a little country store," Tate asked: "'Do you think you can make a machine that will fly?' Picking up a cigar box lid, Wright sailed it several yards and said: 'That piece of wood is heavier than air but, driven by motive power . . . from my wrist, it flies. If some way can be found to give a flat surface sustained motive power . . . then men will fly. This thing of being heavier than air is no obstacle. Birds are heavier than air.'" To Tate, the incident was a revelation.[11]

The Wrights found Kitty Hawk unlike what they had expected. It was a tiny fishing village on a comet-path peninsula streaking north into Tidewater Virginia and trailing sparks of elongated barrier islands as it burned a path between ocean and sounds. Partly hidden in a copse of woods, it was one of several weather-beaten hamlets, three to fifteen miles apart, that were strewn along a 250-mile desolation of mostly treeless, undulating sand barrens, where only ribs of ruined ships broke the featureless shoreline.[12]

Kitty Hawk itself, though, was less a hamlet than a five-square-mile cluster of tiny settlements bonded by dependence on the Sound Landing post office, where Addie Tate, having succeeded her husband, was titular postmistress. The tiny village of about sixty-six families stretched from the Kitty Hawk Lifesaving Station by the ocean and meandered westward along a path through the woods, past settlements at Primitive Baptist Church, Boaz Southern Methodist Church, and Poor Ridge, to Sound Landing.[13]

The path linked families "up the road" (northeast) and "down the road" (southwest). The shelves of Jesse A. Cahoon's small store held little but staples and a few "notions," the livelihood of nearly every family coming from commercial fishing, notably during spring herring and shad runs. Apart from fishing, and hunting wildfowl and razorback pigs, there was little gainful employment. The lifesaving

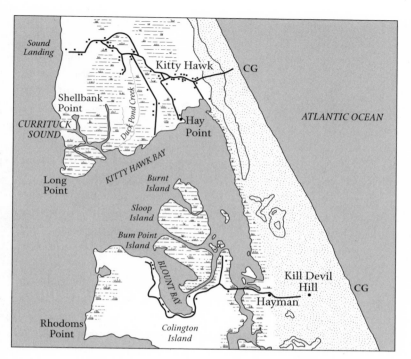

Sound
Landing

Kitty Hawk

CG

Shellbank
Point

*CURRITUCK
SOUND*

*Duck Pond Creek*

*ATLANTIC OCEAN*

Hay
Point

Long
Point

*KITTY HAWK BAY*

*Burnt
Island*

*Sloop
Island*

*Bum Point
Island*

*BLOUNT BAY*

Kill Devil
Hill

CG

Hayman

Rhodoms
Point

*Colington
Island*

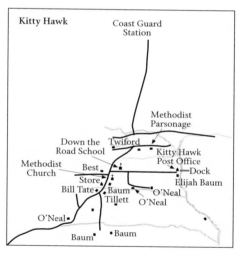

**Kitty Hawk**

Coast Guard
Station

Methodist
Parsonage

Down the
Road School

Twiford

Kitty Hawk
Post Office

Methodist
Church

Best

Dock

Store

Elijah Baum

Bill Tate

Baum
Tillett

O'Neal

O'Neal

O'Neal

Baum

Baum

station housed the weather station, with its telegraph for reporting storms and wrecks to the mainland.[14]

A few, called surfmen, worked year-round at the lifesaving (later Coast Guard) station, where they had two-month vacations, government pay, sturdy cottages for families, and dormitories for the rest. From stations every seven miles or so along the coast, patrolling surfmen warned away unwary ships and aided stranded ones. To save crews, cables were often shot across ships' bows in horrific weather, or rowboats challenged towering waves to reach them.[15] Surfmen's work often was associated with unspeakable danger.

Despite their slivers of arable soil and sometimes harsh weather, the islanders enjoyed a simple abundance of fresh and salt seafood and smoked herring. Most had gardens, fig bushes, a few gaunt cattle, game animals, and a tough-minded resolve to make do with whatever it pleased a benign but inscrutable God to provide. Most drinking water came from rain barrel spigots; at night, dim light emanated from kerosene lamps.[16] Daily life here had scarcely changed in two centuries.

All alike were boat people. Because of the hardship of travel over pathless soft sand, a visit "from people living nine miles from your house [450 miles in the case of the Wrights] was an event . . . long recalled." Dr. John L. Cogswell, an improbable Connecticut Yankee (and Tate's brother-in-law), provided the community's medical care. In a 12-by-16-foot one-room school "down the road," Joseph Cahoon taught a bland but serviceable curriculum of hygiene, math, and geography.[17] In olden days, some Outer Bankers had lived by scavenging beached ships (maybe even luring some to their doom) and others by whaling and other honest pursuits. Many natives, like the Tates, were the progeny of shipwrecked sailors who chose to remain and seek wives among the beach people.[18]

Scores of these tanned, weather-hardened people would come to know the work of the Wrights. Their isolation was a welcome shield for the Dayton men from journalists and curious mainlanders during this and later seasons of activity at Kitty Hawk. "All the men . . . and women," said Tate, helped to "guard their secret."[19] This sensitivity and cooperation was a godsend to the Ohioans, who expected neither.

Kitty Hawk lifesaving crew, 1900. Library of Congress, Washington, D.C.

On a fall morning, Tate and the Wrights took the glider and tent to Lookout Hill, a tall dune half a mile south of town, to begin the experiments. With the glider at first tethered to a jerry-built wooden derrick, Tate watched raptly as the $22.50 contraption wafted from the top of the dune, sometimes ballasted by chains he had lent them. The weather station sent up its handheld anemometer to measure wind velocity. Increasingly, as the months and years passed, the Wrights found themselves dependent on such local assistance.[20]

Tate raced through his own daily chores so he could run over to Lookout Hill and help with the tethered flights. He "worked so hard," insists a contemporary, "that . . . it endangered his health, and some of the . . . family went to the Wrights and said it would be a good thing when the experiments were finished because . . . Bill couldn't stay away from the camp and he was doing two men's jobs." One day he took his sailboat over to Elizabeth City for supplies—a

Bill Tate's home in Kitty Hawk, as renovated before 1912, where Wilbur and Orville Wright stayed during their first weeks on the Outer Banks. The first Wright glider was assembled in the front yard. Courtesy of Daniel Grady Tate.

Bill Tate and his family on the porch of their Kitty Hawk home, which also contained the town's post office, 1900. Bill holds younger daughter Irene; wife Addie holds older daughter Lena. Miss Nancy Baum stands beside Addie. Library of Congress, Washington, D.C.

three-burner stove, dishes, and gasoline. Having never encountered the fuel before, he was alarmed by it: "I never will forget that I wrapped it in a tarpaulin & put it away up forward in the boat so it would be as far from me . . . as possible . . . but I took the wrong precaution. I should have left it open." In any case, he got home safely.[21]

Kitty Hawk as seen from Lookout Hill, 1900. Library of Congress, Washington, D.C.

On October 3, 1900, the trio held the untethered glider by its twenty-foot ropes and ran downhill until it began to rise. Wilbur hopped to a prone position on the lower wing, but, unable to steady it, had to be pulled down quickly, his first glider flight an unnerving failure. The apparatus was not achieving the lift that the Wrights had expected.[22]

Islanders found the glider tests even more entertaining than the glider building, and word spread. On October 17 Tate's stepbrother Dan Tate and Dan's eleven-year-old son Tom stopped by. Dan helped out, and Tom was engaged as seventy pounds of human ballast in some tethered flights, the second person to fly a Wright glider—and more than a year before Orville's first flight![23]

A still-taller dune stood farther down the beach, and the Wrights were eager to try it for better downhill speed and lift. On the eighteenth, Bill's old horse, "Don Quixote," carted glider and tent three miles to 105-foot Kill Devil (or Run) Hill, the tallest dune on the Outer Banks, within a mile of Lifesaving Station Number Thirteen. There the Wrights set up the big tent, and Wilbur was soon attempting free flights. But the glider still did not behave well and the flights were short.[24] Moreover, some days brought howling winds, others dead calm, so delays were frequent; more time was lost carrying Bill Tate's boiled well water four miles daily. If the brothers returned here, they would need a hangar and a nearby well. Though disappointing, the glider flights had furnished them with a wealth of data that would enable them, back in Dayton, to make major improvements on their next glider.

As the Wrights left Kill Devil Hill on October 23, the men heaved the battered glider from the top of the dune into the sand below. Tate seized the chance to buy the gasoline stove, a marvel to the

Dan Tate, the Wrights' only paid laborer during
their first three seasons on the Outer Banks.
Courtesy of Daniel Grady Tate.

Outer Bankers, for three dollars, though he seems never to have marshaled the will to use it. The Wrights also bequeathed him the wrecked glider, asking only that he forbid photographs and not discuss it with strangers. Pleased with their seaside romp, they headed that afternoon for Elizabeth City in Dan Tate's sailboat.[25] The Banker folk were sorry to see them go.

## THE CONJUREMEN

Kitty Hawkers had expected almost any day to hear tragic news from Kill Devil Hill. They tell the "Morticians Tale" of a man who drove a wagon each flying day to the edge of the camp, parking until the day's tests ended. He then returned to Kitty Hawk, "shaking his head solemnly." He was "the undertaker and the wagon . . . a hearse." The concern registered in the unlikely tale, in any case, was genuine.[26]

At first, said surfman William S. Dough, people were "awful sus-

Tom Tate, the son of Dan, with fish he has caught and the Wright glider behind him. Tom flew on the glider in 1900, more than a year before Orville Wright. Library of Congress, Washington, D.C.

picious." Strangers were rare and, "when the Wrights came . . . with boxes and machinery and whatnot," rumors "travelled down the beach.'" Some thought them "'spies,'" a few "that some powerful 'conjure' was in preparation"—as, indeed, it was. They proposed "driving the makers from the Beach or even lynching them." Local folks, says Bill Tate, "believed in one Good God, a bad Devil, and a hot Hell, and that . . . God did not intend man should ever fly."[27]

The beach people, therefore, were initially reserved toward the oddly accented strangers and their absurd kite but gradually began to respect them. Bill Tate was struck by their frugality, though "they lived well. I knew because I . . . lunched with them at the tent often." Will Dough thought them "'Gentlemen, those boys. Never a hard word, never an unkind act. All the people learned to love them.'"[28]

Surfman John T. Daniels especially admired their work ethic. The brothers arose every morning at 7:00 and worked through the day, stopping at 5:00. "I never saw two men so wrapped up in their work," said Daniels. "They had their whole heart and soul in what they were doing, and when they were working we could . . . stand right over them and they wouldn't pay any more attention to us than if we weren't there. . . . After their day's work" they were "the nicest fellows you ever saw." Bill Tate recalled hearing Wilbur "tell a carpenter who was helping him . . . that 'a nail dropped was not worth the time it took to pick it up.' "[29]

In the memory of Tate's daughter Lena, the Wright brothers were "very nice to children, and . . . brought up in a Christian way of life. When they came to our house, they were very considerate of mother, carried up their own washwater and did everything for themselves." In Lena's watchful eyes, Wilbur was "the leading spirit . . . , and Orville followed." Elijah Baum was impressed that they "never worked on Sundays [or attended church services] but didn't mind if a few of us—they didn't like crowds—visited the camp."[30]

Bill Tate also appreciated them for having "no superior airs" and being "the soul of courtesy. . . . The rule of 'after you my dear Alphonse' was held as rigid as iron by both. . . . Approach a gate or door and there was no chance that you could arrive . . . first. . . . Courtesy and deference to all alike" was their guide. "They liked telling stories" that "carried a moral or [were] intended to make a man think for himself."[31]

To Lillie Etheridge, wife of surfman Adam Etheridge, they "were nice looking men" but mingled little. "They just . . . kept to themselves," going only to the [Kill Devil Hill] lifesaving station for the mail. "I never saw them go inside; they would just go to the door, and someone would take their mail to them. It wasn't that they didn't like people, . . . just very secretive." But Bill Tate recalls that "[t]hey talked to us about the things they thought we were interested in," and anyone seeking advice always had "the sympathetic ear of the Wrights."[32] The Ohioans, in short, gradually opened themselves to the Outer Bankers' undemanding hospitality.

Bill Tate was anxious to foster the brothers' return but unsure how to go about it. In November he informed them that the Outer Banks had enjoyed an "Indian summer," since "the day you departed." Nets were taking "plenty of nice ocean fish."[33] His point seemed to be that they could expect better conditions the next time they came.

Curious about the glider, Tate wrote the Wrights of going with neighbor Walter W. Best to retrieve it, clearing away drifted sand, and carting it home. Having "dissected its anatomy," he found it "builded better than I expected." He was making implements from the wood, and his wife Addie sewed wing-cotton dresses for their daughters, aged three and two.[34]

Tate was a rock of support throughout the year. His letters, addressed to Wilbur as "Dear Mr. Wright," were invariably obliging and full of advice about the wooden shed they wanted for their return visit in 1901. He sent them Elizabeth City lumber prices, judging rightly that they needed "simply a rough house rain proof with floor." He recommended vertical siding, the boards "not Lapped & the cracks covered with strips, to keep out sand." A well sunk anywhere nearby would yield good water and local labor, of course, was readily available. He would help as often as possible.[35]

Tate also proposed that they ask the owner's permission to build at Kill Devil Hill, a request that Captain Daniel Webster Hayman would probably allow for "a mere pittance." Asked to discuss this

**Kitty Hawk Lifesaving Station, 1900. Library of Congress, Washington, D.C.**

with Hayman, whose home was on the sound side about half a mile west of the hill, Tate was assured by Hayman that a building might be erected, provided that the Wrights took it down when they were finished with it.[36]

During the summer of 1901, before the Wrights returned, Tate bought "Building Materl & Roofing felt" from Kramer's mill in Elizabeth City and had the lumber "squared and cut" as directed. He recommended that the Ohioans buy their groceries at Elizabeth City from A. F. Tovey and Company, gasoline from N. W. Grandy, and building supplies from Sharber and White, "who treat you fair," plus eight feet of well-driving pipe.[37]

Israel Perry's "Royal Blue passenger service," he noted, was defunct, as a result of bad publicity after Wilbur's harrowing voyage. But the sloop *Oshun* left Kitty Hawk for Elizabeth City on Mondays and returned the next day; if they wired from Norfolk, Tate could meet them at Elizabeth City in his sailboat "any hour you wish." Tate wondered "what the cientifics [*sic*] say" of their experiments and where to find "literature on your work here." The James Hobbses

**Wreck of the 1900 glider, from which Bill and Addie Tate made tools, dresses, and other things. Library of Congress, Washington, D.C.**

sent their regards and a reminder that the Wrights owed them thirty cents.[38] Tate's reward for his service was his pleasure in giving it.

On July 11, 1901, the Wrights reached Kitty Hawk with a larger glider and stayed overnight at the Tate home. The next day Bill carted their crates to Kill Devil Hill and helped erect the tent. To ensure their right to build, Wilbur and Orville walked over to visit Captain Hayman. The grizzled great-grandson of a derelict Englishman, Hayman was surprised by their request: "'Well,'" he said, "'there's lots of room out on the beach. . . . The neighbors let their cattle graze on my land—but I don't care. I would not want to drive the beasts to other pasture. . . . What'll you do with the land?'" "'We want it for experiments in flying,'" one of the brothers replied. The old captain "burst into loud laughter. . . . 'Flying experiments? Ha! ha! a dog's hind leg—fly in the air! . . . Boys, go down on the beach and use it all you like.'" But the Wrights insisted on a formal agreement, so the site was "leased for the munificent sum of $1.50" for the season. Carpenter Oliver O'Neal was hired to build a workshop–living quarters and Dan Tate as a day laborer, a position he held for over two years.[39]

The 1901 season was abysmal, the mosquitoes arrayed in brigades of winged piranhas to guard their beach from interlopers. On July 26 Wilbur tried nine times before getting the machine up, and then for just seconds. The new glider misbehaved in crosswinds and could not make turns. After each try, noted John Daniels, they "discussed its behavior" and launched it again. "Then they sat down . . . and entered notes in a little book."[40]

Dismayed by bad weather and poor performances, the Wrights packed the glider in the shed on August 20 and left, after less than six weeks, unsure if they would return. Wilbur, for the moment, doubted that humans would ever fly. But the beach had revealed its unparalleled potential as an incubator for flying experiments, and they would certainly be back with another machine. In the Kill Devil Hill surfmen they had found an indispensable source of free labor and companionship, and they were adapting well to local deprivations. Since they were coming back, it was all right with Hayman if they left their shed standing.[41]

The brothers had long since found that glider wings must twist to offset tilting in a gust, causing the glider to dive out of control. Wilbur's wing-warping system, twisting the wings in flight to rectify the imbalance, worked well. Still, the plane sometimes slid toward the lower wing, driving it into the sand. They needed to find a way to prevent that or give up flying.[42]

The flights, meanwhile, were having a powerful effect on the Outer Bankers. It was rare that anything so rewarding to watch occurred on these forbidding shores, and the lamp of convivial debate and speculation flickered through many an evening. A glimpse of the glider, bearing a man aloft on invisible shafts of air, was a sight long to be cherished. Most Bankers felt that the Wrights were chasing phantoms but would welcome them back in 1902.

On nonflying days, natives were riveted by the sight of the Wrights imitating wildfowl. The beach teemed with gulls, gannets, geese, hawks, and other birds of many sorts and sizes. Unexpectedly, the brothers had found that Kill Devil Hill was a marvelous natural laboratory of precious data on the secrets of how winds and wings interact. Close study on idle days repaid the experimenters handsomely by pointing the way to better flying machines.

"The Wrights killed sea gulls," explained Will Dough, "and studied how the wings flapped and the bones were put together. Yes, sir, . . . , they would . . . watch the flying, soaring, dipping and swooping gulls. For hours and hours. We didn't know what to make of it. . . . It was wonderful the way they could imitate those big wings. Bending and twisting their wrists, and flapping their elbows just like the gulls."[43] John Daniels said that they seemed most interested "in gannets . . . big birds with a wing span of five or six feet." They imitated "with their arms and hands, . . . every movement of the wings." Gannets, alternating rapid wing beats with gliding and diving for fish with wings half closed, became models for various design changes. The Ohioans set bird traps and dissected hawks to compare the dimensions and proportions of their wings with those of the glider.[44]

Studying fish hawks showed that, when the wind was calm, they lifted "straight up . . . in such a way as to leave a break in the air, a vacuum, discernible to the eye." That vacuum, if it could be mechanically induced, could enhance the machine's lift. The Wrights also "spent a lot of time . . . on top of the hill," observed another Outer Banker, tossing bits of paper "and watching them sail off." To Lillie Etheridge, "[t]hey looked just like little children playing, they did."[45]

According to Archie Daniels, son of surfman John Daniels, the Wrights "watched every kind of bird in that part of the state . . . trying to figure which could go the longest without flapping its wings, soar the highest, dive, bank and turn." Settling on turkey buzzards, "they pulled out a .22 rifle and shot one down." They "took over 200 measurements of that bird, from every possible angle."[46] In this way, Orville evolved an idea that "soaring birds while floating . . . with motionless wings, keep their equilibrium by adjusting wing-feathers at different angles to meet shifting winds, each feather independently controlled." It might be possible to build a plane that, like such birds, could hang in the air "with a stalled motor," while its pilot "revels in the scenery below."[47] The Wrights quietly began to consider establishing a permanent facility at Kill Devil Hill.

Despite their keen interest, the surfmen in 1901 still saw little to suggest that powered flight was possible. In 1928 surfman T. N. Sanderson reminded Orville that he used to come to the lifesaving office, "and I told you not to waste . . . time" flying, it would "never amount to anything."[48]

Striking his usual upbeat tone, Bill Tate wrote in October 1901 of fine weather and a blessed absence of mosquitoes: "I only hope if you ever come . . . you may strike 4 weeks like the past 2 and if you dont fly under those conditions you had better give up."[49] The weather was always perversely superb, the winds impertinently steady, after the Wrights had left.

The brothers' optimism revived in the winter as they conducted wind tunnel tests and perfected their design. Hearing in January of an Atlantic storm, they asked Tate about damage to their camp. He "peeked in through a crack, [and] the old Machine looked like it was enjoying a long rest." He thanked Wilbur for copies of an address he had given to engineers in Chicago and hoped to see them soon in the camp. In July he wrote of dry weather and "the fewest mosquitoes ever known, a happy comparison with last summer."[50]

## THE ORACLE

The fliers were back on August 29, 1902, on the *Lou Willis*. Deck-hand Truxton Midgett, hinting at the general view that their project was futile, asked pointedly if they actually expected their new glider to fly. "Truxton," he was told, "we don't know. . . . But we've studied it, and even if we die trying, we know the principle is sound." They hoped that this craft, with its thin wing surfaces and vertical-vaned rear rudder, would outperform its predecessors.[51]

Getting to camp was aways an ordeal, but this time more than usual. Dan Tate ferried the Wrights and their gear from Dosher's wharf in Kitty Hawk to Baum's wharf, on the sound near Kill Devil Hill, where they hired a horse and cart. Jesse J. Baum, who was thirteen years old in 1902, later remembered his father's order to "hitch up Bold'un and I went to the pasture, caught our big, black horse, and hitched him to the cart." Struggling across the dunes, the Wrights had to push the cart "all the way over one hill." Jesse, lingering while they opened their crates, was amazed by "dozens of sizes of small nuts, screws and bolts and other equipment."[52]

Will Dough recalled that the Wrights, needing help to repair storm damage, build a hangar, and perform other chores, called on Captain Jesse E. Ward at the lifesaving station "and said they would like to have the help of his men in their experiments. The captain said 'yes' and allowed us to run over to the Wright camp whenever we were off duty. . . . Oh, it was lots of fun for us. . . . I have a fortune in memories. I'm glad I helped."[53] Others felt the same way. Rookie surfman John Daniels, a strapping Roanoke Islander, was especially useful, his five-foot-ten, 240-pound frame a solid daily presence. "Every day we were free," he said, he and his brother-in-law, Adam Etheridge, "would be over there helping. Everything we did was free gratis."[54]

In tests of the new glider, Wilbur thrilled onlookers by easily making shallow turns, a payoff for months of hard work. By this time an expert, he taught Orville to fly and, within days, they were each gliding five hundred feet or more. The new rudder allowed them to increase wind drag on either wing, as necessary, to restore balance.[55] But crashes were still frequent. One sleepless night Orville speculated that, by making the vertical rudder movable, the pilot might turn it toward the high wing and rectify balance. The next morning Wilbur agreed, adding that the wing-warping lever should be wired to the rudder so that both responded to one command. This symbiosis of insights, typical of the Wright brothers, was a brilliant step toward mastering controlled flight.[56]

"They . . . got their glider working," said Daniels, "so they could jump into a wind off that hill and stay in the air for several minutes, gradually gliding down to the beach almost as graceful as a gannet could have done it." He and his friends were beginning to appreciate that the Wrights "were a long way from being fools."[57]

Many Outer Bankers were alarmed that the fliers' presence drove up food prices. At Theo Meekins's Manteo store they periodically bought beans, apples, canned goods, eggs, salt, and salt pork. Their bill was small, he ventured, because they "just didn't have the money" and "denied themselves all luxuries and many . . . necessities" in sacrifice to their goal. Bill Tate estimated their food cost at five to eight dollars a month, including eggs from housewives and orders to Elizabeth City. One day, asked by a curious youth why they kept so many eggs on hand, Wilbur replied playfully: "See that little black hen over there? She lays six eggs every day." Though Orville was a fine chef, said Dough, they dined simply.[58]

That fall the Kill Devil Hill activity, so far unknown to the world press, caught the attention of Walter L. Cahoon, editor of the *Elizabeth City Tar Heel*. He had probably just met Lorin Wright, an older brother of the airmen, who passed through town on September 29 headed to the camp. Four days later the *Tar Heel* carried a remarkable article, the first public notice of what the Wrights were doing. Its prescient title was "Men Will Fly."[59]

Cahoon revealed that next season they would "attach an electric motor and propeller . . . and vie with the birds in flight, defying gravity and adverse winds." He knew of the beach machine shop where they were "thinking, planning and perfecting . . . transportation through space." The "aeroplane," he had been told, "is a flimsy boxkite affair with a number of silken wings and steering attachment of like material. With this frail contrivance the aeronaut leaps from the crest of the highest sand dune . . . and, borne upon the wind, he steers his queer craft right or left, goes downward or upward until the velocity of the wind abates and compels a descent."[60]

The experiments, he continued, "have proved successful and the enthusiastic inventors will next summer attach propelling power . . . and no longer rely upon the wind. They have discovered that their machine will work . . . as long as the wind maintains a velocity of eleven miles an hour. They have perfected a motor and propeller to drive the machine and all [that] remains is to connect the two." This was the fullest and most accurate report on the Wrights prior to 1904. Though carried soon afterward in the *Raleigh News and Observer*, it seems to have been otherwise unnoticed and, to the brothers' satisfaction, generated no further comment.[61] Even in Elizabeth City, where the Wrights sometimes appeared when material and supplies were loaded on the freight boat, they were ignored. (Its first automobile, excitement enough for the moment, had arrived in July.) In September a sound-side excursion steamer had come in close to the camp, perhaps so passengers might glimpse the gliding.[62]

As before, in 1902 local help was a requisite for the Wrights. Surfmen sent over catches of bluefish or spot as often as their duties allowed. They helped carry the glider up the slippery dune after each trial and ran errands for tools, food, building materials, newspapers, coal oil, and so on.[63] Camp mail, delivered thrice weekly to the lifesaving station, was brought over on ponies by Daniels or Etheridge. Other Kitty Hawkers gave timely advisories of approaching weather changes and helped repair storm damage. Lillie Etheridge repaired some of the frequent tears in wing cloth on her sewing machine, and Bill Tate oversaw the hauling of lumber, recruited men to build the hangar, and ferried the brothers to Elizabeth City as the need arose.[64]

The islanders also provided the Wrights with almost the only society they knew on the Outer Banks. A revered occasion was a dinner at Bill Tate's, where the main course was wild goose, a dish unfamiliar to the Ohioans. Tate warned them of its "wild gamey flavor," which some found unpleasant. His day was made when Wilbur asked: "Please pass the platter . . . , I want some more of that wild gamey flavor." For the Wrights, it was a welcome change from a diet overreliant on fish. They also sometimes dined with the Etheridges and the crew at the lifesaving station.[65]

The airmen were now competent and daring enough to compete with each other "in seeing who could best glide the machine down the hill" and "alight . . . at the foot." Getting up speed, wrote Tate, "they would level out the machine at the foot of the hill, sail over level ground, and, as the speed slackened, alight on their feet without either wingtip touching the ground."[66] Appreciating the skill they had gained, Tate's patience wore thin: "[W]hy in hell don't you put an engine on it and *fly* it." Daniels concurred. "One day," he said, "they flew for more than 600 feet against [*sic*] the wind, gaining altitude as they went, until the contraption was set down voluntarily." By the end of October they had made over one thousand flights, no longer running downhill and jumping on but lying "prone to start with" and being "launched by two men, one at the end of either wing."[67]

Every flying day seemed to bring new adventures. Daniels liked to tell of a flight when he became "tangled in a rope and hung on for all he was worth. . . . 'I was afraid my time had come,'" but, as the glider descended, he was able to slip off unhurt. On another occasion, the machine spooked Etheridge's horse, which "pulled the stake up and ran away," leaving its rider to walk back to the station.[68]

On October 28, ready to go home and build a motored plane, the Wrights stowed the glider in its hangar, got Dan Tate and James Hobbs to cart their baggage to Kitty Hawk, and left. The 1902 season had been a watershed, convincing them that they now knew how to fly an airplane.[69] At least a few of the islanders were coming to relish the prospect.

Wilbur wrote Bill Tate in the spring of 1903 to ask that he get a permit from Captain Hayman to build a new shed. Tate replied, as usual, that he would do so, but he doubted that the owner would be "as enthusiastic as you suggested about the immortalization of their soil, & the Cash donation Biz." Rather, Hayman was likely to seek as much as five dollars this time. (The "Cash donation Biz." is obscure; Wilbur perhaps joked that Hayman should pay him to use the dune or lower the rent since Kill Devil Hill was soon to become famous and increase in value.)[70]

In the summer Tate wrote to remind the brothers to buy a barrel of gasoline at Standard Oil in Norfolk. He had his own gasoline launch and was now employed by "a Northern family" at Martin's Point, five miles north of Kitty Hawk. Burdened by his new job, he would see little of them that season but would try to attend to any business they wanted.[71]

## THE SOUNDING OF THE REED

When the Wright brothers arrived in September 1903 on the tugboat *Guide*, they had with them a big 625-pound glider, propellers, and a gasoline motor. A few days later they were up in the old glider and, on November 5, began using the new *Wright Flyer*, not yet mounted with its motor.[72] Succeed or fail, the beachmen expected the 1903 season to be an epoch in their lives.

Local interest inspired a growing list of visitors: Bill Tate, Kitty Hawk farmer George W. Twyford, fisherman Walter Best, twelve-year-old John Collins, John Moss, a Mr. Hollowell, and a Reverend Davis, besides Dan Tate, the ever-present surfmen, and others. The camp took on a tone of high-spirited camaraderie. Sometimes in the evening or on Sunday, the Wrights were invited to Kitty Hawk for amateur theatricals or other events or to circulate as guests; they often invited visitors and helpers to dine with them at the camp. A popular pastime was target shooting with .22 rifles, Daniels being reputedly the best marksman of the lot.[73]

Following a November hurricane, Dan Tate reported six ships beached nearby. "Mountainous seas" had left the schooner *Mabel Rose* stranded to the north, the *J. H. Holder* a wreck to the south. But two days later, nets at Corolla, near the Virginia line, landed thirty thousand storm-driven mullet, the biggest catch in memory. Dan, his head turned by visions of cash to be made at the nets, demanded and got a raise to seven dollars weekly. Nevertheless, he left two weeks later "to take charge of a fishing crew."[74]

The Wrights were ready to test the *Flyer* with its motor. On the beach one early December morning, they came upon Alpheus Drinkwater, an ebullient telegrapher from the Currituck Beach Lifesaving Station up the coast, who had come down to splice a broken wire. "'Flown yet?'" he asked. "'Not yet,'" said Orville. "'The hurricane . . . tore up our camp, and that gale the other day ripped it apart again. . . . [B]ut we'll be trying it soon. We'll let everybody . . . know when we're ready. I hope you can come.'"[75]

Drinkwater, who devoted his time to matters of consequence, thanked him but needed to be where the real action was. His telegrams brought tugs from Norfolk to save ships grounded on shoals. He must watch for "big news breaks" and could ill-afford "'to spend much time away down here.'" "'I see,'" said Orville. "'Well, good luck.'"[76] (This is Drinkwater's account, but inviting a telegrapher from thirty-five miles away is unlikely to have occurred to the Wrights. Will Dough contends that they never issued "official invitations. . . . They just talked it around, you know.")[77]

In late November the motor was installed and, according to Daniels, the "sprocket chains rigged from it to the propellers." In ground trials, backfire broke a propeller shaft, which was shipped express to Dayton for repair. If Samuel Langley succeeded in his scheduled trial flight in Washington on December 8, the glory would be his and the Wrights would be merely a footnote in history. The shaft was soon fixed but cracked again, and Orville rushed home for a new one. He was back on December 11 with gratifying word that Langley had failed.[78]

At 1:30 P.M. on December 14, the surfmen saw a red flag at the Wrights' camp summoning them. Most of the crew was thirty miles north at Wash Woods, where, on December 3, one of the navy's two submarines, the *Moccasin*, had beached after breaking a tugboat hawser at the mouth of the Chesapeake Bay. Drinkwater was eagerly tapping out wires. The surfmen had seen many flights, but a submarine was altogether a novelty. Ironically, then, it was the *Moccasin* that thrust the Outer Banks into the limelight in December 1903. Before then, stray rumors of flying being done there had not been widely credited.[79]

Those responding to the flag were John Daniels, Will Dough, William Thomas "Tom" Beacham, Bob Westcott, Benjamin W. "Uncle Benny" O'Neal, and two boys drawn by the noise and stir. This crew, says an Outer Banker, helped lay a sixty-foot monorail of spliced two-by-four scantlings, "resting on posts one and one-half feet high," down the northern slope. They carried the airplane to the top and put it on the truck, an eight-foot-long pallet with a pair of in-line, "double-flanged" bicycle hubs and ball-bearing axles to roll it along the track.[80]

Daniels recalled the Wrights having a brief chat: "I saw them flip a coin to see which one would go first." Wilbur chose heads and won. He walked over and lay down in what Outer Bankers referred to as the "Belly Bumpums" posture on the lower wing and, trembling with the cold and anticipation, called for the restraining wire to be clipped.[81] Orville and Daniels ran alongside for about forty feet to hold the wings and keep the plane from tipping. "It climbed a few feet," said Daniels, but "settled to the ground near the foot of the hill. . . . The left wing touched first. The machine swung around and dug its skids into the sand, breaking one. . . . Several other parts were also broken." His stopwatch showed that it was aloft just three seconds; it had risen too fast owing to the dune's nine-degree slope. The Wrights, unfazed, vowed to try next from a level surface so that skeptics could not claim that they had only glided downhill.[82]

The next day the plane was repaired, and the Wrights relaid the

The *Wright Flyer* on Big Kill Devil Hill prior to the unsuccessful trial on December 14, 1903. Library of Congress, Washington, D.C.

monorail at the foot of the dune, but the day was windless. They resolved to try again on the eighteenth, but the seventeenth opened with an unexpectedly brisk twenty-mile-an-hour breeze, so the red flag went up again around 10:00 A.M.[83]

Daniels—just back from patrol—Dough, and Etheridge were disappointed at missing the submarine but trudged over. With them were lumber buyer William Cephus "Ceef" Brinkley, who had come up from Roanoke Island for shipwreck timber, and seventeen-year-old John T. "Johnny" Moore. Johnny, from Nags Head Woods, was only passing by but "decided to stick around . . . to see the show" and help lift the plane onto the truck. Six-year-old Lloyd Dough, Will's son, scampered off when one of the Wrights playfully said "'C'mon, we'll take you up.'" Young John Ward of Nags Head Woods evidently fled with him.[84]

The brothers, said Daniels, "got their machine out . . . , and we helped them roll it out to the foot of big hill."[85] Only the Wrights seemed to be aware that a transcendent moment in history was at hand. Now and then, the men huddled at the stove to warm their hands against the bite of the chill breeze. Some wild razorback pigs were chased off so they would not interfere with the flight. The brothers "walked off from us," continued Daniels, "and stood close together . . . , talking low to each other for some time. After a while they shook hands, and we couldn't help noticing how they held to each other's hand like they hated to let up, like two folks parting who weren't sure they'd ever see each other again." "Good luck," said Dough. "The Lord knows," Orville replied.[86]

Wilbur led Daniels over to a tripod Korona V box camera near the right side of the plane and asked him to put the black cloth over his head and take a picture as the machine left the track. Daniels, chosen for his marksmanship, had never taken a photograph but was told that it was "like sighting a rifle." "Look in the window . . . and when it rises . . . just squeeze this thing here."[87]

"It had just rained and a stiff wind had frozen the puddles into solid ice," said Dough; "we tried our best to cheer Orville. We clapped our hands, but," in the cold wind, "we didn't get up much enthusiasm." "There was considerable tension," noted Daniels, no one "felt like talking." The motor was warmed up. Wilbur removed the bench from under the right wing that kept the plane from tilting. Orville lay down on the wing, shifted the gear, and signaled for the wire to be clipped.[88]

Johnny Moore recalled that men were positioned at each wing tip, Wilbur on the right and Etheridge on the left, as the plane "moved down the . . . rail increasing speed as it went." "My place," said Dough, "was at the tail. . . . I held on until she pulled loose." It was 10:35 and the wind was up to twenty-seven miles an hour.[89]

The plane went off "with a rush," said Daniels, and "left the rail as pretty as you please," though its propellers were turning so slowly that the revolutions could be counted. Daniels squinted into the camera and, at the instant when the plane was "all in the view-finder . . . squeezed the bulb."[90]

For all of its timeless significance, the flight had about it the

John T. Daniels's photograph of Orville Wright in the first powered flight, December 17, 1903. Library of Congress, Washington, D.C.

appearance of fumbling ineptitude. Daniels noted that the "front rudder was balanced too near the center," and the plane tended "to turn itself." It would rise to "about 10 feet and just as suddenly aim toward the ground." The trouble, he supposed, was that Orville "pulled her nose up too high and she began to slip. Then he dropped her and managed to level out, but couldn't maintain his altitude" and came down.[91]

Aesthetically, the event was so far inferior to many of the graceful glider flights that it struck Johnny Moore as a bungled effort. One had to realize, as he and perhaps others did not, that a machine, for the first time in history, had flown into the wind, not on it. The wonder of the flight could not be detected with the eye alone. The primordial pull of gravity for the first time loosened its grip on a heavier-than-air machine; universal law itself seemed to have undergone fundamental alteration.

Orville was able to cover 120 feet in twelve seconds (a good runner could have done it in five), before "a wing tilted," Daniels added, "and caught the sand. . . . We cheered him . . . but there were six of us and we didn't make much noise." "None of the five Banker witnesses," notes another, "remembered a word . . . any of them said to anybody."[92] In the still of the moment, however, the soft notes wafting eastward over the sand might have been those of an ancient reed.

Witnesses later offered contradictory impressions of the first flight. Daniels was the one person present with enough of the poet in him to invest the moment with passion: "I don't think I ever saw a prettier sight in my life. Its wings were braced with new and shining copper piano wires. The sun was shining bright that morning and the wires just blazed in the sunlight like gold. The machine looked like some big, graceful golden bird sailing off into the wind, . . . it made us feel kind o' meek and prayerful like." (On less inspired days, Daniels conceded that "we were used to seeing the glider . . . , so we weren't much impressed.")[93]

Johnny Moore, spry but vacuous, and having seen better flights, left, "rowed his skiff back across the creek and thought no more about it." Over a third of a century after the event, said an interviewer, it did not seem to Moore that "anything out of the ordinary" had occurred that morning. Decades later, when the *Wright Flyer* was installed at the Smithsonian, he was asked why he did not glance up at the suspended plane: "I seen it onct," he replied. (Yet a son of his, born on December 17, 1929, was named Orville.)[94]

The Wrights flew all morning. On the second try at 11:05, according to Daniels, Wilbur meant "to fly around the coast guard station and back, but he wasn't flying much higher than my head and finally his rudder got caught in the sand and pulled him down." Orville, on the third flight, hit a knoll and "the rudder came off. I was standing on the lee side, . . . jumped and tried to pull her down. . . . I was lifted off the ground . . . and fell back into a sand dune."[95]

On the fourth and last flight, Wilbur proposed to fly all the way to Kitty Hawk, a feat that would garnish the day with many other witnesses. He prudently rejected the notion but stayed in the air fifty-nine seconds and covered 852 feet—a distance about equal to that of the mouth of Salmon Creek. It was the only flight not marred by an accident, but only minutes passed before another occurred.[96]

**The Kill Devil Hill lifesaving crew, including those who helped launch the
December 17, 1903, flight.** *Left to right*: **Captain Jesse Ward, William Thomas
Beacham, Adam Etheridge, John T. Daniels, and William S. Dough. Courtesy
of Outer Banks History Center, Manteo.**

Daniels, accident-prone and seated on a wing while the Wrights
were "tinkering with a fitting" and considering a fifth effort, told the
tale. A thirty-five-mile gust suddenly "swept across the beach just
like you've seen an umbrella turned inside out." Wilbur hopped
nimbly off, but Orville and Daniels held on until Orville, who was
"as fast as lightning and quick as a cat climbed right through . . . and
came out the side." Daniels, bigger and slower, did not make it.[97]

"I got tangled up," he explained, "in the wires that held the thing
together." Others grasped at the tilting plane, Wilbur seizing the
rear uprights to no avail. "I found myself caught in them wires,"
Daniels continued, "the machine blowing across the beach, heading
for the ocean . . . rolling over and over and me getting more tangled
up in it all the time. . . . When the thing did stop . . . , I nearly broke
up every wire and upright getting out." Daniels would always take
pride in being called the first airplane casualty.[98]

Not until the Wrights reached home and developed their photographs did they see that Daniels's picture of the first takeoff had perfectly captured the unforgettable moment. It was the most tangible evidence they would ever have that what they claimed was true. In the photo, the airplane is suspended about three feet above the end of its track, just above the truck. Orville lies prone on the lower wing, and Wilbur, arms akimbo, stares intently from the right, his back to the camera. Beyond lies a boundless, featureless sand and skyscape summoning Orville toward a limitless, unfathomable future. Even so, the photo revealed entirely too much about the plane's construction for the Wrights to allow it to be published or passed around. Over a quarter century had elapsed before Orville, in 1930, mailed Daniels a copy, so that he could view at last the result of his marksmanship.[99]

Outer Bankers recalled this epoch as one of uncommon interest and pleasure. They best recalled amusing stories and picturesque episodes. Yes, a plane had flown into the wind, but it was not so much the technological triumph that stirred them as their participation in a merry exploit. To them, the Wright era on the Outer Banks would always be remembered as one of good yarns, fond memories, and—oh, yes—the first powered flight. (At his first sight of the *Wright Flyer*, Johnny Moore had asked what it was. Daniels explained that it was a "duck-snarer," which would drop nets over flocks of floating ducks to ensnare them.)[100]

What does not appear in the Wrights' published reflections is the extent to which the beachmen contributed to their achievement. They had summoned the fliers, recommended a site, leased it to them on nominal terms, and befriended them without reservation. They kept them posted on sea and weather conditions, performed the heaviest labor, dug wells, kept them supplied with food and materials, built and repaired huts, and shielded their privacy. The Outer Bankers had assisted in all of the glider and airplane tests, kept off-season vigil over camp property, hosted the Wrights at meals and overnight at the lifesaving station, advised them on building construction, delivered coal oil and milk, and a great deal more. In the end, Adam Etheridge spoke for them all: "I never enjoyed anything in my life as much as watching those boys fly. And I watched them a lot."[101]

In 1928 the issue was briefly raised whether the monument soon to rise at Kill Devil Hill might be dedicated, in part, to "the coast guardsmen who gave so unselfishly of their time in helping the Wrights." Some favored "a combination coast guard station and aviation beacon."[102] But the moment passed and the surfmen were forgotten. None of them complained that they had received no public credit. Yet it was not just technological expertise that allowed the Wrights to fly, but the gladly given help of scores of willing sons and daughters of the sand.

# The Demon of Hill Thirteen

### THE SECRET OF THE DUNES

If destiny decreed that history's first airplane flight would be performed at Kill Devil Hill, did it also will that North Carolina would soon afterward become a major focus of the aviation industry? The answer is clearly no, since the state lacked the industrial base and financial strength to nourish a sophisticated new industry—or so it seemed in 1903. But the Sorcerer, by definition, was a maker of miracles, and the brothers Wright, in fact, were thinking along precisely this line.

Delayed by storms, the Wrights left Kitty Hawk on December 22, 1903, on the schooner *Nancy Hall* and were home by Christmas Eve. Dayton's newspaper, like those elsewhere, initially rejected as a hoax the news of their success. Only the *Norfolk Virginian-Pilot*, informed of the first powered flight within about ninety minutes after it was made, carried a version of the story the next morning, December 18. The *Pilot*'s account, largely fabricated and wildly inaccurate, was still a classic scoop in journalistic annals, by the same paper that, eight months before, had scorned the *Charlotte Observer* for believing that "We will yet fly."[1]

Just how the *Pilot* obtained its story is often debated. But after the second flight at 11:05 A.M., it appears that Adam Etheridge and John T. Daniels rode briefly over to the lifesaving station. There they wired Harry P. Moore, a *Virginian-Pilot* reporter to whom they had promised to send word of any powered flights: "Wrights flew in motor-driven plane 11:20." (They may have considered the first flight a failure.) Norfolk telegrapher C. C. Grant relayed the message to Moore. At about 3:00 P.M., Kitty Hawk weatherman Joe Dosher tapped out Orville's triumphant wire to the Wright family in Dayton. Grant again informed Moore, who then telephoned the Kill Devil Hill Lifesaving Station and received from those present a merrily inflated account.[2]

At a Kitty Hawk meeting with Moore in 1928, Orville Wright confirmed that "'when you called the coast guard station . . . about our flight, the station man turned to me and asked me whether they should tell you. . . . I told them to tell you nothing, but in their enthusiasm they did give out the story and made it a bit stronger than it was. I believe they told you we flew three miles.'"[3] The lifesavers may have supposed that such a tale, though incredible on its face, would help safeguard the Wrights' secrets.

In the days after the first powered flight, many papers carried the news but usually only as a brief Associated Press dispatch. Editors routinely dismissed claims of aerial prowess, but, as some of the facts came out, aviation men from near and far sent the Wrights congratulations and requested details of their machine. The Wright brothers, anxious to retain a monopoly on their technical advances, rejected all such appeals. (Their secrets soon leaked out anyway.) Meanwhile, editors of other newspapers praised the *Charlotte Observer* for insisting all along that men would fly—perhaps, in its way, as gratifying an outcome as the *Pilot's* scoop.[4]

Bill Tate, after giving so much of himself to the Wrights, could not be consoled that he was absent when they flew, having assumed that December 17 was too windy for flying. Writing to Wilbur on the twenty-sixth, he explained that he had gone to Elizabeth City on business. He alluded to the *Pilot* article of December 18 but seemed unaware of its errors, though he had had over a week to contact the witnesses.[5]

Despite his absence on the fateful day, Tate was in a position to cash in on his intimacy with the Wrights and their work. He had received a telegram from the *New York Journal* offering "good pay" for a six-hundred-word article. The *Norfolk Dispatch* spoke for five hundred words, Tate replying, probably not having seen the December 18 *Pilot* story, that the "[e]xperiment was conducted Very secretly and nothing will be learned." In his letter to Wilbur, Tate said he hoped that the flights "were satisfactory" and apologized that Dan played "the fool & left you. . . . If I can be of service to you any time let me know."[6]

The Wrights, no longer dependent on winds, hills, or treeless sand, were now content to make their experiments on a field near Dayton. There, they were seldom noticed, with the result that some newspapers grew even more skeptical about their work. Most Americans remained unpersuaded that the Wrights had solved key problems of heavier-than-air powered flight, let alone that, month by month, they were improving their planes and flying skills. In 1904 the brothers built an aircraft that made flights of up to a mile. A third, in 1905, was the most successful yet; by October they could keep it in the air for up to forty minutes. They added a seat for the pilot, who could lie "Belly Bumpums" only so long, and increased their ability to control the machine. By the year's end, they were trolling for governments interested in buying planes, a search hindered by their refusal to give demonstrations, allow inspections, or provide photographs or specifications.[7]

In August 1906 Bill Tate asked them for their Kill Devil Hill buildings and offered to send anything they wanted from them. He and his family, he added wistfully, "often think of you" and "the fun we had as very pleasant recollections." In November Adam Etheridge wrote to the same effect; the buildings, he observed, were sanded up and soon would be useless, but they had good wood. He wished to do some fishing in the spring and use the lumber for a camp of his own. The Wrights informed both men that they would retain the structures for possible further use.[8]

Kitty Hawk and Kill Devil Hill were no longer of interest to the public, even in North Carolina. The Wrights perhaps gone forever, the *Moccasin* refloated, Outer Bankers resumed their old routines and timeless isolation. None saw reason to record his or her experiences during the 1900–1903 period, and no reporter or curiosity seeker came to learn more. It still seemed unlikely that airplanes would be of much consequence to human existence.

By the spring of 1908, with no orders for planes in hand and rivals Tom Baldwin and Glenn Curtiss making their first flights in New York, the Wright brothers began to fear losing their lead. Trajan Vuia, Alberto Santos-Dumont, and J. C. R. Ellehammer had made short flights in Europe in 1906, Henri Farman and others in 1907. Many surmised that the Wrights' claims had been groundless, while other aeronauts were demonstrating publicly what they could do. The Wrights, in fact, had not flown since 1905 and then essentially in secret.[9]

Wilbur and Orville emerged from their cocoon in early 1908 for negotiations with the French and American governments, both of which demanded demonstrations before making offers. With so much depending on the upcoming trials, the airmen decided to return briefly to Kitty Hawk to test a new steering device. On April 6, 1908, Wilbur set off, again ahead of Orville and a crated plane.[10]

The Wrights wrote in advance to Bill Tate for an update on the condition of the camp and received his reply just before leaving. Writing with his old enthusiasm, Tate said that he had followed their progress in the *Literary Digest* and other publications. The camp was blown down and its contents were gone, but he had a new site in mind for them eight miles up the beach—close to his current residence. A triweekly motor launch ran between Elizabeth City and Kitty Hawk, but he now owned "the best and fastest launch in N.C." and could meet them in Elizabeth City. Updating them on local news, Bill reported that Tom Tate was living with him, as his father Dan had died in 1905. Bill's daughters Irene and Lena, now ten and eleven, sent regards, as did his wife Addie. The James Hobbses were "old and feeble."[11]

On April 9 Wilbur stopped in Elizabeth City to order three thousand board feet of siding from Kramer Brothers, some hardware from G. S. White and Company, and gasoline from Grandy, all to be sent to Kitty Hawk via the *Lou Willis*. Three days later he arrived at Kill Devil Hill in Captain Franklin H. Midgett's gasoline launch, the *B. M. Van Dusen*.[12]

The Kill Devil camp looked as though it had been the object of a mortal assault. The oldest of the two camp buildings was roofless, a foot of sand and debris on its floor. The newer one had disappeared altogether in a recent storm. What remained of the 1902 glider was covered with sand. Moreover, the site had been pretty thoroughly looted by a group of boys from Edenton, among others.[13]

Anxious, since his time was limited, to get his lumber promptly, Wilbur, on Saturday, April 11, walked the five miles to Kitty Hawk and asked Captain Midgett if it could be delivered on Monday. Midgett protested that it was Easter weekend and he could not deliver before Tuesday, but Wilbur clarified the situation by offering him twenty dollars, instead of the regular fee of six dollars, to get the lumber to him by Monday morning.[14]

"Now that $20.00 offer," said Truxton Midgett, "spoke to us. . . . We were to have a special service at our church the next day and . . . got under way as soon as possible. We had a good run . . . and arrived in Elizabeth City that [Saturday] afternoon. We left town the next morning at 3 o'clock and made it home in time for Easter service. We delivered the lumber Monday A.M. . . . and we all felt good."[15]

Carpenter Oliver O'Neal put up the new building, but it was slow work due to the difficulty of transporting lumber from Kitty Hawk to Baum's wharf. Wilbur slept and ate with the surfmen until the camp was ready to be occupied, Orville finally arriving with the plane on April 25.[16]

A new problem was the presence of a *New York Herald* reporter. He disclosed on May 6 that Orville, flying that day for the first time in three years, had a fine hop of 1,008 feet. There were more flights on May 7, the longest for fifty-nine and a half seconds and 2,230 feet. More journalists arrived, but the Wrights, refusing to fly whenever they came within a mile of camp, banished them to some woods at Hayman's Bay, over on the sound.[17] The *Virginian-Pilot*, despite the stature it had gained with its 1903 scoop, sent no one, printing only what it learned from other papers.

Wilbur and Orville were now gaining renown, and the dunes people treated them with renewed respect. The *Charlotte Observer* reported that "people of that section . . . take much interest" in the new machine and "are aiding the Wrights in every way possible" — as, of course, they always had.[18]

With a reporter of the *London Daily Mail* and two from *Collier's* magazine present, the press witnessed what it could of several flights on May 11, including one by Orville of over 8,200 feet. John Daniels was amused that the reporters hid in bushes a mile away and concocted exclusive "interviews with the Wrights." "But I reckon," he concluded, "they had to earn their money somehow."[19]

The 1908 flights answered all skeptics and showed that the Wrights had, after all, solved the mystery of flying. On May 14 Wilbur flew for over two and a half miles, taking with him fellow gliderman Charlie Furnas, America's first airplane passenger. Telegrapher Alpheus Drinkwater sent out "42,000 words of news stories," though, he later claimed, he "still didn't believe a word of it." The *New York Herald* initially demanded that its man stop sending, as it thought, nonsense, and confine himself to facts.[20]

On the afternoon of May 11, Wilbur set off on what he meant to be a flight of about eight miles among the dunes. He had been in the air for about seven minutes when reporters saw the plane disappear behind a dune and heard the engine cut off. They learned that night that he had crashed, though the accident was not serious; neither he nor his machine were badly scathed. Even so, Wilbur, pleased with his plane but plagued by the hot weather, now broke off the trials to address the next business—trips by him to France, and Orville to Fort Myer, Virginia, for army trials.[21]

Walter L. Cahoon's *Elizabeth City Tar Heel*, the nearest paper to the scene but relying on distant reports, was soon describing imaginary Kill Devil Hill flights of eighteen to thirty-two miles. The Wrights, wrote Cahoon more plausibly, had halted the trials only because of the presence of reporters, after smashing all speed, distance, and duration records. They were said to be already negotiating a sale with a Mr. Wu of the Chinese government.[22]

With few exceptions, North Carolina press coverage ranged from ludicrous to nonexistent, though several papers had contacts on the scene who sent occasional reports. The *Raleigh News and Observer* and *Wilmington Morning Star* were among the larger newspapers to disregard the flights altogether. The *Greensboro Patriot* published two brief stories, one claiming that on May 14 the plane flew eight miles under good control but was "an unrecognizable mass" after the crash. The *Raleigh Caucasian* had the Wright brothers flying thirty miles on May 8, with "a fifteen mile dart seaward." The *Concord*

*Times* simply ignored the Wrights in favor of local excitement on the fifteenth, when "Master Freddie," a French balloonist, suffered a nasty fall at the intersection of Means and Union Streets.[23] (A bird in hand was worth two on the sand.)

The *Charlotte Observer*, as harbinger of manned flight, was one of the few North Carolina papers to manifest keen interest in what was going on. On May 5, for example, the *Observer* commented that the fliers were confounding skeptics. There was talk of a possible thirty-mile flight to Oregon Inlet and back "inside an hour," or even to Cape Henry, sixty-five miles north.[24]

Wilbur was soon giving exhibitions at Le Mans and elsewhere in France and dazzling spectators. The French, who considered themselves the anointed of the aviation world, fell hapless victims to his exploits and acknowledged him unreservedly as their master. He set new records almost everywhere and took up about forty passengers. Suddenly, he was a celebrity, the toast of Europe and the Americas.[25]

Meanwhile, with the second machine, Orville was trying to meet the army's performance requirements at Fort Myer. Despite a tragic September 17 wreck that killed his passenger, Lieutenant Tom Selfridge, and seriously injured himself, his flights earned high praise. In 1909, backed by wealthy investors and starting to profit handsomely from contracts, the Wrights formed an aviation company and began manufacturing planes for public and private customers.[26]

In September they heard from Adam Etheridge, who had closed down the camp for them in 1908, again asking for their building. He had recently found that it was broken into and "the box with the machine in it is busted." The brothers turned him down but, for five dollars a year, retained his services to keep the building "in as good order as possible." By May 1910 the roof leaked "and will be bad before summer," but Etheridge could fix it if they liked. He had closed the house, placed a post in front of the door, and nailed the windows shut. "Anything you want," he added, "that I can doo for you just let [me] know." By November, because of his diligent care, the camp was in "first straight condition." He hoped that before long they would fly again at Kitty Hawk. "I always look in the paper for your work & am glad to see what you are dooing." Wilbur replied that they would be back in the summer of 1911.[27]

The new challenge for the Wrights was to find a way to achieve automatic stability, or complete control of their machines, without constantly working levers. The issue was whether a plane could fly level and straight without the pilot's intervention. In 1908 they had found a possible solution, a vane-and-pendulum system that used compressed air to activate wing warping and elevator controls. Wilbur designed a glider version of their 1905 plane for tests, and Orville made plans to return to Kill Devil Hill to try out the system. In mid-September 1911 they sent their brother Lorin to reopen the camp and make other arrangements.[28]

By this time, the Wrights were among several teams around the country hiring and training pilots to give exhibitions at state fairs and similar venues. There was good money to be made this way, and the testing of the planes, as well as the rapidly increasing skill of pilots, helped push aviation forward month by month. Wilbur and Orville were confident of their own expertise but had to find ways to stay ahead of the pack.

Lorin Wright spent the night of September 20, 1911, at the Southern Hotel in Elizabeth City. Reaching camp, he found it again in sad disrepair; it was in worse shape than before because it had now become a tourist attraction and was frequently looted for relics, but principally because the 1908 building was gone. He cleaned up the mess, oversaw repairs, and built a new shed near the old site. He also secured a guard of surfmen to protect the new glider.[29]

Orville headed once more for Kill Devil Hill accompanied by an English friend, aeronaut Alexander Ogilvie, and Lorin's son Horace. In Elizabeth City, they were reported to have a great deal of "paraphernalia, . . . machinery, material and supplies." This time, observed *Elizabeth City Advance* editor Herbert Peele, the Wrights were "the cynosure of all eyes." They were "more communicative now" than in 1908 and "a great deal more sociable, fame and riches having made them magnanimous and charitably inclined." Lorin was "not secretive in the least . . . , making no effort to conceal the purpose of his trip to Kitty Hawk." The Wright party spent several days flying kites and scooting about the sounds in a rented motorboat before

Orville arrived with the glider on October 13. By the time he reached Kitty Hawk, four reporters were waiting for him.[30]

The global celebrity of the Wrights was the carrot that drew North Carolina's press corps from its normal haunts. But the flying would have to compete for public notice with the state fair (Ohio governor Judson Harmon gave the keynote address), the Barnum and Bailey Circus, baseball's World Series, and sundry college football games. Hendersonville's *French Broad Hustler* got the reporting off to an awkward start with a story that the brothers were about to initiate another revolution in flying—a plane that did away with the motor. This gaffe was especially egregious since a glider had recently been performing near Asheville, thirty-three miles north of Hendersonville.[31]

In Raleigh, a chastened *News and Observer* arranged with a Manteo contact to keep it abreast of activity at Kill Devil Hill. Orville caught a boat from Kitty Hawk and reached Colington Bay on Friday the thirteenth, heading for his new campsite near Hill Thirteen, within a mile of Kill Devil Hill's Lifesaving Station Number Thirteen. No one saw an ill omen in any of this, but, as was soon demonstrated, it foreshadowed misfortune for both the Wrights and North Carolina.[32]

After waiting out adverse winds on October 16, Orville had a flight of over four hundred yards; on the eighteenth he twice broke the world glider endurance record by staying up for one and a quarter minutes. On the last of five flights that day, the machine was damaged in falling from fourteen feet, but Wilbur, unhurt, was "delighted" with his day's work.[33]

October 23 began, says the *Charlotte News*, with heavy rains until about 4:00 P.M. Then the sky cleared, so Orville and his crew took the glider halfway up the dune. After four unsatisfactory flights, he noticed that the right rudder's vertical vane needed shifting to the left. From thirty feet up the dune, he made a fifth effort, this time with the rudder corrected. The glider "fairly leapt into the gale" and climbed to about sixty feet. It briefly remained motionless, though "trembling violently." For five seconds the fabric resisted the strong winds before the glider "lurched up and down," bobbing like "a giant box-kite in an erratic wind," and settled to earth. Orville "was jubilant." His machine had, momentarily, hovered almost motionless, like the gannets he had watched for so long.[34]

His pulse racing, he made two more brief flights but, on a third, encountered a harsh gust and began losing control. "The glider strained and lurched . . . and started to buck." Orville desperately worked at the levers but could force the glider no higher than sixty feet. Descending gradually, he was above "a small hummock" when he swerved to the right and began to plummet. According to the *Charlotte Observer*, Wright scrambled for safety as the craft turned and was on top of it when it hit the ground "bottom up." He was unhurt, except for torn trousers and "a slight shaking up." Lorin and Alexander Ogilvie ran to the plane and found the right wing "reduced to kindling . . . hanging . . . by the canvas covering." Etheridge, a witness to the crash, later declared that it was the only time he ever feared for Orville's life. But the damage was not as bad as it seemed then, and Orville was flying again within a day or two.[35]

Trying to improve performance, the crew enlarged the rudder and lengthened the tail for better control. As a result, October 24, 1911, was one of early aviation's most memorable days. With a fifty-mile-an-hour gale and word of a hurricane passing over Cuba, Orville took the glider up twenty times, each flight longer than the

Kitty Hawk landscape, 1911. Library of Congress, Washington, D.C.

one before. Gale-driven sand pelted him with "tiny particles cutting like small shot," virtually the worst possible conditions for glider flying.[36]

Reaching fifty feet on one of his flights, he held his altitude for an almost unbelievable nine minutes, forty-five seconds, his machine adjusting by itself to every alteration of wind, scarcely moving as it hovered. It was the most stunning glider flight ever yet made, a world record that stood for a decade. It showed, the News and Observer mused, that "aviation depends almost completely on a greater knowledge of . . . air currents and greater skill" in adjusting to them. It was a culminating moment in twelve years of experiment and study. It would not be long, hazarded the Spencer Crescent, before "the hum of the aeroplane may be heard at night and seen in the day time as they pass over our homes, our factories and our farms" carrying mail, passengers, and "the commerce of the country at a very rapid rate. Such sights will be commonplace before our children are grey headed."[37]

More flights followed on October 25 and 26. Perhaps woozy with success, Orville talked seriously of staying for the approach of the hurricane. He wanted to try the glider in high winds to give it "a more severe test than it has yet undergone." Luckily, the storm veered toward Mexico, and coastal warnings were withdrawn.[38]

This time few of the Wrights' former Outer Banks helpers, except for the surfmen, took part in the work. Bill Tate brought his family down briefly, and George Baum of Kitty Hawk paid a visit. Orville had his own crew and needed little other help; the glider weighed only 145 pounds and was easily carried up the dune, so that some ninety flights were made during the brief season. But the Kill Devil Hill lifesavers continued to be of invaluable service. Captain Jesse Ward, visiting Elizabeth City in late October, spoke eloquently about Orville flying "to beat the band." Indeed, wrote editor Peele yet more expansively, Ward was "the Wrights right hand man" and "knows more about their machine than any other man."[39]

Meanwhile, a nasty imbroglio was unfolding at Kill Devil Hill. In mid-August 1911, before Orville left Dayton for the Outer Banks, he had received a troubling message from Adam Etheridge. It said that Kill Devil Hill surfman Bob Westcott was tearing down the 1905 Wright building. Orville immediately wired back that "[i]f Mr. Westcott has torn down any part of [the] shed please request him to replace it at once." After speaking with Westcott, Etheridge told Wright that the surfman "said he was going to continue taking it down." Most of it, in fact, had already been hauled to a place near the lifesaving station "to build him [Westcott] a dwelling house." Westcott, talking "as if he knew what he could doo," told Etheridge that "it was on his land and he had a perfect rite to it after it had been vacated."[40]

Etheridge was in an uncomfortable position. Orville had finally given the wreck of the 1908 shed to him and John Daniels, who used the lumber in barns they were building on their Roanoke Island farms. He informed Orville that, as a member of Westcott's lifesaving crew, he could ill-afford to "get Crossed up much whare we are at, here together." Yet "any thing I can doo for you let me know—I will surley doo it." Only now, it seemed, did Orville learn that Westcott had apparently leased the Kill Devil Hill tract from the Haymans—perhaps for the express purpose of acquiring the building.[41]

Here the matter remained for several weeks, but there was much more at stake than the shed. As far back as late 1903, the Wrights had planned to build a large, permanent facility at Kill Devil Hill. As first conceived, they would, in the spring of 1904, erect "more commodious quarters," one that would contain "an extensive and complete shop . . . with all the necessary tools and appliances for building our machines."[42] It seemed clear that they intended to live and work much of that and subsequent years at Kill Devil Hill.

Nothing more, however, was heard of the project for several years. Then, before Orville reached the Outer Banks in October 1911, he was reported to have bought "a large area of land at Kill Devil Hill" and had the parts for "three buzzard-shaped machines" should "the first one meet with an accident."[43] This was exaggerated but not materially different from Orville's plans for the near future.

The Wrights, by this time financially secure, wanted to abandon the drudgery of business and devote themselves to full-time experimentation. As Wilbur wrote a friend in early 1912, they "wished to be free from cares so that we could give our own time to . . . the science and art of aviation"; they had been "compelled to spend our time on business" for the last five years.[44] The facility that they intended to establish would vault North Carolina to the forefront of world aeronautics.

There was an obstacle, however, to implementing the plan. Lorin Wright, after reaching Kill Devil Hill in September 1911, seems to have leased the John J. Mann tract and built the new shed. This land abutted the Kill Devil Hill tract's northern boundary and was only a quarter of a mile from the base of the dune. Thus Lorin may have unwittingly erected the new building within the boundaries of Westcott's lease.[45]

The Wrights were soon at odds with Westcott, not only over the fate of the old building but also, it appears, over their application to him to use Kill Devil Hill for more experiments. An October 13 report mentions a misunderstanding "between Wright and Bob Wescott [*sic*], owner [*sic*] of the land" used in earlier tests, "for which reason it . . . may interfere with the inventors plans."[46]

The Wright lease from Mann embodied an option to buy his tract. The property contained three large dunes; one of them, the 100-foot Hill Thirteen, was almost as high as Kill Devil Hill itself, and would be available as a backup site if the old hill could not be used. But Westcott finally agreed to let the brothers use Kill Devil for their experiments,[47] perhaps at a suitable fee. On leaving the Outer Banks in late October, Orville felt certain that he would soon exercise his option to buy the Mann tract and, in the spring of 1912, build on it a permanent laboratory-hangar in the heart of his seaside aviary.

In late November 1911 the *New Bern Daily Journal* carried a lengthy report from a Manteo source who knew Orville's plans in considerable detail. The anonymous letter stated that, on November 23, John

Mann had received word that the Wrights would exercise their option to buy his tract "and establish a permanent experiment station" on it. The one thousand acres included "giant sand dunes" and woodland stretching west across the Outer Banks from the ocean to Croatan Bay. During the winter and spring, the Wrights would erect at Hill Thirteen "a combination hangar and laboratory of concrete" and, possibly, "a bungaloe for living purposes . . . , the only structure of its kind in Dare County."[48]

The new building, "for the permanent testing of aeroplanes," was to be "much larger than the old [1911] building and will be thoroughly equipped for aeroplane building." It would provide protection from any kind of weather, be wide enough to hold two flying machines, and "in all probability a set forge will be installed at one end." The structure would also have a carpenter shop and storage loft. The Manteo writer speculated that the Wrights would continue their arrangement with Westcott (or Hayman?) for the use of Kill Devil Hill when any gliding was needed.[49]

Although situated in a remote region, guaranteeing the brothers a measure of privacy, the writer pointed out, the Wright experimental station would have easy communication with the world beyond. It was within ten miles of a Manteo telegraph office, six miles from the Kitty Hawk post office, where they would post and receive mail, and one mile from the telephone line at Kill Devil Hill Lifesaving Station. The railroad at Elizabeth City was forty miles away, but a daily packet ran between there and Manteo, besides a thrice-weekly boat from Elizabeth City to Kitty Hawk.[50]

The Wright-Westcott unpleasantness involved a possible error by Lorin regarding the Mann-Westcott property line. A hidden complication was that the dune, which shifted gradually southward on prevailing winds, was now about seventy-five feet from its 1903 location. Lorin, less familiar than Orville with the terrain, may for this reason have unwittingly put the building on Westcott's side of the line, making it subject to his control. (Lorin was reported to have had the 1911 shed built "at the foot" of Kill Devil Hill.) It must have seemed in Dayton that the brothers dared not launch their laboratory project until the issue was firmly resolved, perhaps by their purchase of Kill Devil Hill when Westcott's lease expired.[51]

In early 1912 the Wrights still had an opportunity to buy Mann's land, build the laboratory on it, and, if necessary, use Hill Thirteen for later tests. But the timing was terrible. Preoccupied with a complicated lawsuit to protect their 1906 patent rights, the brothers may have concluded that there was no point in trying to unravel the mess until other, more pressing matters were resolved. At any rate, the dream vanished forever on May 30, 1912, with Wilbur's untimely death from his old nemesis, typhoid fever. Soon afterward Orville retired from the airplane business.[52]

Part of the Wright legacy to North Carolina was that, by 1911, Tar Heel artisans were designing or testing many different flying machines of their own. A permanent Wright presence on the Outer Banks could only have intensified such efforts across the state. Thus Wilbur's death apparently denied North Carolina a chance to become a focal point in global aeronautics and aviation on the mainland as well as the Outer Banks. No such opportunity would arise again in the twentieth century.

In November 1912 Adam Etheridge sent Lorin Wright a plaintive appeal expressing sorrow at Wilbur's death and hoping that he and Orville might eventually "fly some around Kill Devil Hill again we have not had such an excitement since you all left." But Adam did not see Orville again for many years.[53]

Apart from their stunning success in aeronautics, their several visits to the Outer Banks had done wonders for the brothers' health and stamina and, it appears, their relationships with other people. Lured by neighborly favors and hospitality from their hideaway amid the dunes, they had grown, as the *Elizabeth City Advance* put it, "more amiable, magnanimous, and charitable."[54] They had also gained maturity, self-confidence, and a measure of robustness by learning to live in their rugged camp apart from the amenities of the life they knew in Dayton. They seem to have left the Outer Banks better men, in many respects, than when they arrived.

It would be gratifying to show how the Wrights publicly acknowledged the help and sustenance—material and emotional—that they received from the Outer Bankers in these years. Apart, however, from an occasional reference to the five first-flight witnesses, they said nothing in print regarding the beach people. Because the Wrights' own published accounts bear narrowly on technical challenges and achievements, their work at Kill Devil Hill has been made to seem that of two men working alone in a forlorn "Arabia Deserta." That unintended version of history, still held almost universally, is, as this book seeks to show, grotesquely misleading.

# The Early Whirlies

WHEN WINDMILLS FLY

That airplanes, like gooney birds, required large cleared fields to land and take off and could not, like hummingbirds, hover in flight, were not initially considered serious limitations. That they could not maneuver in tight spaces or, if desired, proceed very slowly across the sky were other restrictions that were ignored in the first thrill of powered flight. But some experimenters had sought all along to build machines with just these capabilities. The promise of powered flight, of the Conjurer's decree, would not be fully realized without them.

The earliest description of a toy windmill, ancestor to the helicopter, dates from at least the first century A.D., when Heron of Alexandria described such a device in his *Pneumatica*, a work of popular science. But it was six more centuries before the earliest full-sized windmills began harnessing the power of the wind for useful purposes. Meanwhile, in the fourth century a Chinese savant, Ko Hung, wrote of a vision of "flying cars with rotating [windmill-like] blades" as future man-carrying machines.[1]

In the West, toys based on the idea of rotating wings—pinwheels, pull-string gadgets that energized rotors, and so on—began to appear in the fourteenth century, toy flying helicopters in the nineteenth. In 1878 the Wright brothers' father gave them a wind-up toy helicopter of French design, which introduced them to the notion of mechanical flight.[2]

As in the case of so many other ingenious mechanisms, it was Leonardo da Vinci who, in 1493, sketched the first man-carrying helicopter. His design featured a spiral screw of starched linen rimmed with iron. A platform below would carry a crew of four, who, holding onto a stationary bar, would treadmill the floor around in unison, producing up to two horsepower.[3] It was an intriguing concept but too low-powered, as Leonardo perhaps understood, to be workable.

An obvious need in any sort of heavier-than-air flying machine was a lightweight power source, and the Age of Steam seemed, for a time, likely to provide it. English experimenter John Stringfellow, who moved to New Jersey in 1849, later produced a sixteen-pound steam engine that generated one horsepower, using distilled wood alcohol for fuel. He caused a model rotary plane to lift off at London's Crystal Palace in 1868, but its power was far less than what a true helicopter would require.[4]

With the appearance of Daimler's internal combustion motor in Germany in 1885, soon mass-produced, an engine light and powerful enough for heavier-than-air craft was at last available. But no aerial experimenter appears to have recognized it for some time. As E. K. Liberatore puts it, "the only wisdom" so far gained "was that nothing tried works," not only in power plants but also in most other technology related to helicopters. Because experimenters worked individually and at random, no body of shared insights was accumulated.[5]

As the twentieth century began, the state of helicopter progress was about what it had been in Leonardo's time, though Enrico Forlanini in Italy coaxed a model steam-driven helicopter into a flight of twenty seconds in 1877. Seven years later, Horatio Phillips in England built a steam model that rose 250 feet.[6] But steam toys were of marginal relevance to the dream of vertical flight for humans. Urged on by the success of their fixed-wing colleagues and advances in gasoline engines, helicopter experimenters made a frontal assault on the problem in 1907. In that year several laid claim to having built the first full-sized rotary machine to lift itself off the ground.[7]

The Breguet brothers sometimes receive credit for the first helicopter flight, which they claimed to have made at Douai, France, in September 1907. Four helpers were said to have held the machine, as it rose a few feet, unmanned and tethered, in order to steady it, since it had no stabilizing controls. But at least two of them later declared that they were pushing the machine up, not holding it down. The Breguets are famous for their aeronautical achievements, but the first helicopter flight probably should not be among them.[8]

Another claimant for the honor of achieving the first helicopter liftoff was Paul Cornu. He is said to have caused his machine, on August 31, 1907, to perform a shallow, unmanned liftoff at the village of Coquainvilliers, near the English Channel. The following spring he published in the French journal *L'Aerophile* an account of this flight and a manned flight that he said he made in November. But there was no witness to either flight and no way to corroborate his claims. On such testimony, any number of people may be said to have preceded the Wright brothers in heavier-than-air flight. Even if Cornu's story were true, however, his exploit, like that of the Breguets, may have come too late to be called *first*.[9]

The distinction of achieving the first helicopter liftoff appears, rather, to belong to an unsung craftsman in rural North Carolina. His name was William Luther Paul.

## THE FLIGHT OF THE BUMBLE BEE

The facts of Luther Paul's life are well known because he was locally celebrated as a versatile inventor and gifted "Mr. Fixit." He was born in 1869 in the isolated Carteret County fishing hamlet of Davis, on Core Sound, just over two hundred miles down the coast from Kitty Hawk.[10]

Paul received little formal schooling but early showed an uncanny ability to mend tools and machines, even those with which he was unfamiliar. He fashioned his own otter and mink traps, fire and oyster tongs, and laborsaving farm and garden equipment. While helping his father build a chimney, he worked out a block-and-tackle elevator for lifting bricks. He built his own bicycle, later rigging its sprocket to a grindstone, which enabled him to perform in seconds the tedium of sharpening knives and axes.[11] As with the cycling Wright brothers and other aviation pioneers, parallels between mastering balance on bicycles and stability in flight may have played a part in expanding Paul's horizons.

After building and operating blacksmith and machine shops, Paul in 1906 installed an inboard motor in a flat-bottomed boat and set out to bring the world to Davis's doorstep. He drilled holes in the sides, rigged a motion picture projector with a gasoline projection lamp, and created a double-side-wheeler "showboat," probably the first waterborne movie theater. Before most cities had movie houses, he was showing films to isolated populations along neighboring shores. He also erected at Davis a merry-go-round, which revolved to such tunes as "Come Josephine in My Flying Machine," and a show house lit by his home-built dynamo. His home-to-shop telephone was the only instrument of its kind in the area.[12]

Luther Paul's love affair with propeller-driven machines may have begun with his "sneak-boat," a turn-of-the-century one-person craft that floated low in the water. Its operator lay prone so that he could come close to floating ducks without driving them off. A small paddle wheel at the rear was turned by the occupant's feet. Paul made several sneak-boats for himself and friends, and local waterfowl became sitting ducks for hunters.[13]

In a 1913 letter to Thomas Edison, Paul outlined an ingenious means of synchronizing sound and film in motion pictures and asked the "Wizard of Menlo Park" to critique it. Edison, perhaps too preoccupied to read the letter closely, responded that he had "successfully completed the [same] invention," which "is now in actual operation in about 40 cities." This was his silent vitagraph system, well short of what Luther proposed.[14] Partly because Davis was a lonesome backwater, talking pictures had to wait another fourteen years.

It might have been more productive for Paul to have sent an account of his helicopter, another field in which Edison was a pioneer. In 1880 Edison proposed that an engine generating three horsepower per pound would be able to lift a helicopter. But his trials of a lightweight engine led to an explosion that almost cost his life and

William Luther Paul.
Courtesy of Grayden M. Paul.

turned him in other, less hazardous directions. He was sure, however, that flying would have little practical application "until someone invents a machine that can stop in midair and hover like a bird and then land . . . anywhere it wants."[15]

Another interest of Luther Paul was windmills. They had been in use in coastal North Carolina since the eighteenth century for grinding corn, drawing water, and other purposes. Although steam engines began replacing windmills in the late nineteenth century, scores of the old type were still in use in coastal counties as the twentieth century opened. Paul built his to power the mill he used to saw boards for a home he was building for his family.[16]

North Carolina windmills were of the European horizontal shaft type, with gears to regulate the turning force of the wind. They had auxiliary tail wheels or directing vanes to help keep them facing the wind. Such mills, usually about forty feet tall, could produce up to fourteen horsepower in a not-uncommon thirty-mile wind.[17] It did not require a trained engineer to appreciate that wind wheels, laid horizontally, might be constructed in such a way as to give them lifting power. It is likely that Paul's belief that he might fly came, in part, from his work with windmills, as well as experience with his paddle-wheel steamboat, sneak-boat, bicycle, and rotary grindstone. He envisioned a plane that rotated its wings for lift.

By his own account, Paul's flying machine "was being tested before Wright flew at Kitty Hawk" in 1903.[18] In any case, there is no indication that the Wrights were his initial inspiration. But the revelation that they had proved mechanical flight possible, and on the North Carolina coast, was a strong incentive to persevere in his aerial experiments.

In addition, an English translation of Jules Verne's *Master of the World* appeared in 1904. It is unlikely that a copy of this tale of a powerful helicopter built in North Carolina's mountains found its way to Paul's bookshelf. But the novel showed that the concept of the helicopter was penetrating the popular imagination and could be expected to influence more experimenters. In five years after the publication of *Master of the World*, at least fifteen helicopters (a figure larger than that for the whole nineteenth century) were either tested, patented, or reported to be under construction.[19]

Paul, seemingly the least well positioned of these rivals to succeed, named his experimental machine the "Bumble Bee" because it could "turn or dodge as quick as a bumble bee" and, compared with airplanes, was "what the Bumble Bee is to the Butterfly."[20] Calculating the desired ratio of weight to power, he fashioned a power plant comprising four motorcycle engines linked together so as to yield about fourteen horsepower. At the top of the frame were two tandem four-blade rotors eight feet in diameter and probably of tube construction. A compound helicopter, one having a wooden tractor-type (pulling) propeller for forward flight, it had horizontal rudders operating in the rotor's wake, a rear elevator, and side vanes for lateral control. Four automobile wheels served as a steerable landing gear. The Bumble Bee could be driven as an automobile or, properly sealed, on water as a boat.[21]

Theoretically, a novice could fly it. The propeller blades, Paul

**A turn-of-the-century windmill near Beaufort. Paul built such a mill at Davis prior to assembling his helicopter, a technology that evolved from windmills. North Carolina Division of Archives and History, Raleigh.**

wrote, were "so anchored in the Hub that a hand lever sets them at zero [angle,] head or back. [The] Propeller is hooked up direct to the motor, but when set at zero [thrust for vertical flight, it] has very little resistance against the motor. You . . . release the clutch that connects the motor to [the] rotor shaft[.] You then start the motor set at zero. When ready to go up you put in clutch and rotor starts. Speed up motor gradually until machine leaves the ground."[22]

At a desired altitude, the pilot set the tractor propeller's pitch for flight in the direction he chose, altering pitch by moving the steering wheel forward or back to control the rudder. Stability was maintained by moving the pilot's shoulder harness, fastened to the side vanes, in the direction opposite the craft's list. He changed direction "simply by turning the steering wheel in the direction you want to go." Thrust was varied by speed change rather than pitch control; vertical descent only required slowing the motor. An umbrella parachute could be deployed to ease the Bumble Bee down in case of engine failure or other mishap.[23] (The notion of a safety parachute for flying machines, still alive in aviation circles, may have been another "first" for Paul.)

He and several coworkers built a large hangar. They left out a back wall so that, with front doors open to the sea breeze, it served as a wind tunnel in simulating forward flight, though perhaps at some cost to the machine's controllability. Tests "by the rude equipment to be had at that early date" showed that the engine produced too little power to lift the machine's weight. Paul sought to improve its performance by polishing the air-intake ducts and filing off as much engine metal as possible.[24] He probably also shed the four auto wheels or perhaps replaced them with skids.

At an unspecified date in 1907, with the Bumble Bee's weight reduced to 500 pounds, Paul claimed, he got it—while tethered and unmanned—to lift between four and five feet inside his hangar. There appear to have been several witnesses to this and other Bumble Bee takeoffs. At the peak of his success, the craft cleared the ground carrying up to 60 pounds of sandbags. The inventor planned to fly in it himself when it could lift his weight—120 pounds.[25]

Luther later confided to his granddaughter Lina that he wished he had "taken the machine out when they had it lifting 60 pounds." He was confident that, "with a good breeze and a hefty push," it would have flown as high and far as the Wrights' craft. On another occasion, he wrote that the machine's weight was within twenty pounds of his goal when be abandoned the project.[26]

But abandon it he did. Returning home one day from work, he found his wife Lina in tears and holding a recent issue of their subscription newspaper, the triweekly *New York World*. It contained a front-page feature and photograph of a recent plane crash, presumably that of Orville Wright at Fort Myer, Virginia, on September 17, 1908. Wright was seriously hurt and his passenger killed in this, the first fatal accident in powered, heavier-than-air aviation. Lina begged Luther to give up his experiment at once. With about six years and a great deal of labor invested in it, he might have resisted her pleas, but other factors weighed in her favor. A cousin in Washington, North Carolina, who had promised to help finance Paul's work, had recently died without leaving such a provision in his will. Also, there had been a disconcerting accident at the hangar only a day or so before when a blade broke loose from a spinning rotor and hurtled through the building's wall "like a .22 bullet."[27]

Paul reluctantly agreed to dismantle his promising machine. In about three years he moved with his family to the nearby and larger town of Beaufort and opened a successful motion picture theater. In 1913 he constructed an automobile, one of Beaufort's first. Still inventing new gimcracks, he came up with a kerosene burner for two-cycle engines, a coin counter to speed up a tedious task at the theater, and a fourteen-record sound effects system that synchronized music and sound with his films.[28]

On America's entry into World War II, Paul discerned new vistas for his old Bumble Bee design. In February 1942 he contacted his congressman, Graham A. Barden of New Bern, and learned that he might bring his helicopter to the attention of the Inventors Council of the U.S. Department of Commerce. On March 23 he sent the council a description and sketches of his machine and its operation. He asserted that modern "Rotor Ships, giro, helicopter and Autogiro machines" could not compete with the Bumble Bee "for lightness[,] compactness and maneuverability." A one- or two-seater based on his design, he ventured, could be as small as eight feet wide and sixteen feet long (probably the dimensions of his 1907 craft). Covered with light material, streamlined, and with rotors disengaged, it could be driven by its propeller as an airboat, and on land, with its steerable wheels, as readily as in the air. Thanks to "the development

Sketch of the Bumble Bee by Luther Paul. The Bumble Bee is said to have lifted four to five feet off the ground in 1907, making it probably the world's first helicopter. Courtesy of Grayden M. Paul.

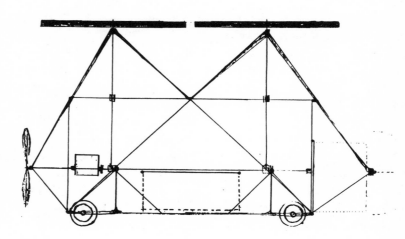

of gas Engines," his machine could be made more compact and agile than any airplane.[29]

But Paul was a war too late. On July 14 J. C. Green, assistant chief engineer of the Inventors Council, informed him that "the suggestion is not of such character as may be employed in the war effort at this time." Green may have doubted that the Bumble Bee could hover without steadying from the ground, and that the use of rotor speed rather than pitch control would mean sluggishness of operation.[30]

The designs that Paul sent to the Inventors Council in 1942, the only known depictions of his helicopter, are not, of course, conclusively identical with that of 1907. But his 1942 sketch of the Bumble Bee's wide rotor blades, its triangular platform, and the placement of its elevator beyond the rudder are features that belonged to turn-of-the-century aeronautics, rather than those of the 1940s. The machine that Paul described for the council was almost certainly the same one that he claimed to have tested "35 years ago."[31]

Paul's matter-of-fact statement that his machine flew in 1907 is, of course, not proof that it did. Several of his offspring (and at least one unrelated witness) have asserted from firsthand knowledge that it lifted four or five feet, but they do not know precisely when. All agree that the liftoffs were achieved at Davis prior to the family's

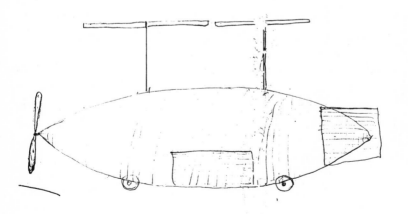

**Paul's sketch of the Bumble Bee as it might be adapted for use on land or water. Courtesy of Grayden M. Paul.**

Model of the Bumble Bee, built by Luther Paul's grandson, Grayden M. Paul, an aeronautical engineer. Courtesy of Grayden M. Paul.

move to Beaufort in 1911 and that work on the Bumble Bee was halted by the *World's* airplane crash story and photograph.[32]

There is presently no way to corroborate the contentions of the Paul family and others and may never be. The earliest-known published reference to a flight by Paul's helicopter appears in a *News and Observer* dispatch from Beaufort in December 1937, when the inventor was sixty-eight years old. Although offering no date for the event, the writer of the story alleges that Paul stopped working on the machine at some point after the Wrights' first flights in 1903. In any event, the family's assertions have been made merely as a matter of filial pride and not as a challenge to competing claims, none of which they seem to have known about until quite recently. Paul's own lifelong reputation for humility and honesty also weighs heavily in his favor.[33]

If, then, the Bumble Bee achieved liftoff in 1907 or, indeed, 1911, it was certainly the first American helicopter and, as seems likely, the world's first. North Carolina, for the present at least, may claim not only the first horizontal flight but the first vertical one as well. Paul's evidence stands on equal or sturdier footing than that of either the Breguets or Paul Cornu, his only rivals.

Luther Paul died in Beaufort on January 11, 1946, leaving five children and an enviable reputation as a resourceful craftsman. The first manned "flyable helicopter," one that solved some of the problems that plagued Paul, is attributed to Russian-born George de Bothezat at Dayton, Ohio, in October 1922.[34]

## THE PHANTOM CHOPPERS

Luther Paul was not the only North Carolinian to work on helicopters in the first years of the twentieth century. The *Wilmington Morning Star* of August 21, 1909, relates that a resident of the town had lately built a model flying machine and that "it flies too. It is heavier than air and when it starts it goes up straight. It requires no skids for it to get right up in the air. Its starting flight is not at a slant, but . . . upward from the ground. The inventor has not built a machine to carry up a man or two, but we would like to see one built according to the model."[35] A rubberband-powered coaxial rotor probably supplied the model's power.

The unnamed builder was almost certainly Frank Herbst, a German immigrant known as the first automobile dealer, beginning in 1903, in eastern North Carolina and perhaps in the state. Around 1908 he began to study aeronautics and became interested in rotary-winged aircraft. He was said to have become "entirely enamored of the . . . thrilling and scientific study of aeronautics." The *Wilmington Morning Star* in 1911 identified him as one of three builders of different "flying machines" in Wilmington at that time.[36]

By 1911 Herbst had become an agent for the Curtiss Aviation Company's barnstorming teams. No flier himself, he arranged many of North Carolina's earliest flying exhibitions. Herbst, who emigrated from Germany at the age of fifteen, also enjoyed later fame as a speedboat builder. Despite the success of his model helicopter, he does not seem to have completed a full-sized machine, and his descendants have no memory of his interest in aviation.[37]

Finally, there is tentative evidence of a helicopter builder in Raleigh at or about the same time. An undated (ca. 1941) memoir of native Wake County machinist Carr E. Booker claims that "[i]n 1904

I was employed as a machinist mechanic. . . . The firm with whom I was employed built for a College Professor, a helicopter of his own design; . . . it was completed . . . by Blue Prints furnished. It did actually lift the pilot about ten feet, [but] as [with] many ventures . . . , finances gave out and . . . further development of this Helicopter ceased."[38]

Since Booker is not known to have worked outside of Wake County, the builder was presumably a professor at present-day North Carolina State University in Raleigh or the University of North Carolina at Chapel Hill. A remote possibility is W. H. Pickering, a Harvard professor, who was working on an electric-powered rotary plane at least as late as 1903. Booker himself built a plane in 1922 and, like Herbst, was an automobile dealer.[39]

The problem with Booker's account is that he was only eleven years old in 1904, when he said the helicopter was built, and most unlikely to have been employed as a machinist or even an apprentice at such a young age. The date may be a misprint for 1914 or an earlier or later year, but the work took place well before Booker's own airplane-building project was launched in 1921. In the meantime, he began selling Crow-Elkhart automobiles in Raleigh.[40]

Because this helicopter's alleged flight was even more impressive, at ten feet high—and manned—than Luther Paul's, it is unfortunate that Booker died before he could be invited to clear up the mystery. No corroborating evidence has yet been found to support his contention. Nevertheless, it appears that North Carolina craftsmen labored at or near the cutting edge of vertical flight in the first years of the twentieth century and lagged behind few, if any, in the world in this challenging new technology.

North Carolina interest in vertical takeoff-and-landing machines soon faded, but it revived half a century later. Beyond the scope of this volume are the "gyrocopters" designed, built, and marketed by aeronautical engineer Igor Bensen in the 1950s and 1960s. Born in Russia in 1919, he became interested in bird flight as a boy and nourished the interest after his arrival in America in 1937. He built his first gyrocopter at Raleigh-Durham Airport in 1955 and, later, versions that could drive on roads and land on snow, ice, or water.[41] In several respects, they were much like Luther Paul's Bumble Bee.

These were one- and two-seater vehicles weighing as little as eighty-five pounds, the smallest up to that time, which were suitable for utilitarian and recreational purposes. They set twelve world records for this type of machine, including four at Kill Devil Hill on December 17, 1955. A flight by Bensen from Raleigh to Winston-Salem, which covered eighty-four miles in ninety minutes at up to 7,400 feet in altitude, was then the longest and highest ever made by such a craft. He built a modest market for his machines and kits before retiring to California after his company folded in 1988. His rotorcraft designs remain an inspiration, especially in the recent mushrooming of interest in sport rotorcraft. Bensen died early in the year 2000.[42]

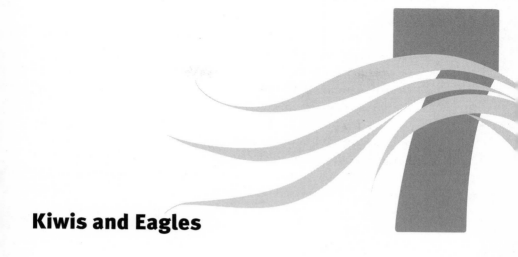

# Kiwis and Eagles

It was in its way a phenomenon unparalleled in techno-logical history. In 1910 and 1911 amateurs all over Europe and America built hundreds of flimsy homemade airplanes and tried, at the risk of their lives, to fly them, sometimes successfully. The catalyst for this burst of creative energy and self-destruction was the impression made on the Western imagination by aeronautical advances in 1908 and 1909. In addition, the first wave of aerial exhibitors appeared in 1910, dazzling thousands who, thirty months before, had not considered such things remotely possible.

Besides those of the Wright brothers, several primitive airplanes achieved flight before 1908 in low, slow, straightaway hops. But in mid-1908 Orville demonstrated his fast and agile new machine for the army, even as Wilbur created a sensation in Europe. Louis Blériot, a round-faced little Frenchman hailed by many as the most daring airman of his time, transformed military and foreign policy precepts on July 25, 1909, by crossing the English Channel in thirty-seven minutes. In September 1910 Peruvian Georges Chavez hurdled the Alps in a flight from France to Italy. And in the fall of 1911 Orville Wright flew again at Kill Devil Hill.[1]

Geographic barriers, it seemed, need no longer inhibit dreams of world empire or sparkling new vistas of commerce and communication, perhaps even the transfiguration of human relationships, self-perceptions, and patterns of thought. Jeering skeptics were transmuted into flying enthusiasts by these and other tokens of the airplane's stunning advance from cricket to Pegasus—all well within the airplane's inaugural decade.

Even denizens of earthbound agricultural regions felt new lightness in their limbs. Many people in the American South, especially North Carolina, by then the region's leading industrial state, were swept up by the vision of soaring majestically above the clouds. North Carolina's prized commodities, wood and cotton, were the raw materials of airplane construction. Few Tar Heels had the training or experience, let alone funds, to build their own flying machines, but they thrived on the faith they saw crystallized in the achievements of others, often with scant resources.

In just twenty-one months, from mid-1910 to early 1912, North Carolina artisans built at least a dozen heavier-than-air craft (not including the Wright planes), most of which actually flew. In at least eleven other instances, Tar Heel projectors of airplanes and helicopters in this period failed at various stages to complete their work. And these were only the ones that came to public attention. All projects so far discovered in North Carolina were in towns possessing at least one daily newspaper. What went on in other places was often only locally known and perhaps soon forgotten.

The nation was experiencing the advent of what Joseph J. Corn has called the age of the "Winged Gospel, America's Romance with Aviation," a love affair that would span almost half a century.[2] During the first half of that era, North Carolinians contributed markedly to kindling this frenzied coupling.

Given the difficulties they faced, North Carolina builders made a surprising and, for them, unprecedented response to the glittering attainments of artisans elsewhere. In a comparable period in the history of the automobile, say 1900 to 1903, only three or four Tar Heels are known to have built horseless carriages, a far less daunting challenge than building flying machines. For over a quarter century after the success of Robert Fulton's *Clermont* in 1807, only two steamboats were built in the state.[3] But airplanes riveted public imagination, churning the wellsprings of creativity as never before.

To some individuals, building airplanes had an irresistible allure. In small towns, the obstacles to doing so were evidently too great to overcome. But this was not true in small cities like Greensboro or Wilmington, Asheville, Winston-Salem, or Charlotte. Here, pooling the expendable assets of a few adventurous businessmen, or a public stock sale, could raise the few thousand dollars needed. A skilled mechanic or an engineer working in a small shop could, with access to the thin body of relevant literature, solve most of the technical problems involved. And success might mean not just the thrill of achievement but, conceivably, fame and riches at the threshold of an exciting frontier of enterprise.

People who thronged to ballparks and racetracks to see professional exhibitions of aircraft soon realized that a brisk breeze or a sudden thunderstorm could ruin everything. Fragile cloth-and-wood planes were hard to control even on splendid flying days, and accidents marred almost every performance. Heavier-than-air flight, plagued by erratic winds and unsuspected air currents, as well as the airman's tentative grasp of the fundamentals, was an exquisitely dangerous and complex activity.

Adequate motors were now available—a secondhand automobile or motorcycle engine might suffice—but aspiring North Carolina aeronauts initially had little luck controlling their machines. Their planes tended to bounce obstinately along fields and beaches or balk at their pilots' commands. Controlling flight remained the key problem to solve, and Tar Heel craftsmen focused their efforts on this goal. Wilmington, the state's first mainland aviation center, provides telling examples.

In July 1910 Wilmingtonians Harold M. Chase and Minor F. H. Gouverneur received a patent for a mechanism to stabilize and control airplanes. Their invention was a manually operated lever cryptically described as capable of being swung in any direction to transmit motion to other elements of the plane, apparently including panels on the wings. Three weeks later they won a second patent for rear wing panels (ailerons) that were controllable in flight by a cable or flexible cord, a device presumably different enough from the 1906 Wright wing-warping patent to discourage prospective lawsuits.[4]

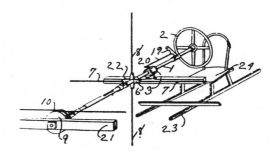

**Chase-Gouverneur 1910 patent for a "controlling mechanism" for airplanes.
Note that the pilot sat in a chair and guided the plane with a steering wheel.
U.S. Patent Office, Washington, D.C.**

Meanwhile, the pair built a plane in "great secrecy" on uninhab-
ited Shell Island on the Atlantic shore, a few miles east of Wilmington. The site was "far up the beach where few people ever go." Chase,
a naval reserve commander, was manager of the American Chemical
and Textile Coloring Company; Gouverneur, an engineer, was general manager of Wilmington's electric street railway and vice president of a cotton mill.[5]

Journalists had no luck trying to catch a glimpse of the *Gouverneur
I*, as it was apparently called by its builders. They learned, however,
that it was a quadriplane: its four huge cloth-covered wings—30 feet
long by 16 feet wide—were arranged in tandem pairs, similar to the
Langley machine of 1903. The aircraft had a 25-foot-long frame made
of aluminum and weighed 1,200 pounds. It seems to have been powered at first with a 250-horsepower motor, a power plant too heavy
for the flimsy structure. The engine cost $4,000, the whole machine
$6,000.[6]

Chase and Gouverneur announced in June 1910 that they would
make test flights at nearby Wrightsville Beach, a summer resort, on
July 4, meaning crowds of witnesses. The plane would fly low "along
the sound side of Wrightsville Beach from Moore's Beach [Shell Is-
land]" to the southern end of Wrightsville Beach. But the inventors,

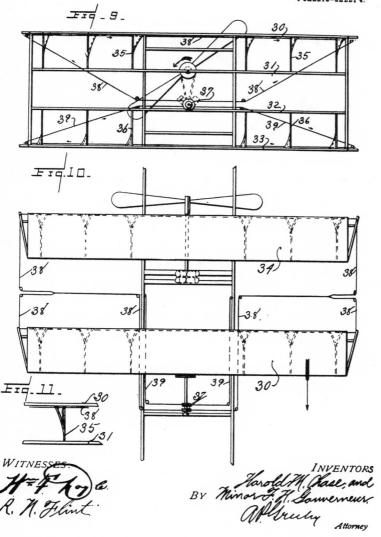

**963,516.**

*Fig. 9.*

*Fig. 10.*

*Fig. 11.*

*WITNESSES.*

*INVENTORS*

Harold M. Chase, and
BY Minor F. H. Gouverneur.

*Attorney*

Chase-Gouverneur 1910 patent for "stabilizing means" for the inventors'
quadriplane, with two wings fore and two aft. Chase flew it at Shell Island Beach,
near Wilmington, on November 16, 1910. U.S. Patent Office, Washington, D.C.

without explanation, abruptly canceled the test, disappointing a throng of five thousand.[7]

The larger engine was replaced with a forty-horsepower Harriman, a big saving in weight, yet comparable in lifting power with other popular aerial motors of the time. On the afternoon of November 10, Chase flew this (or possibly a second) plane in a series of unannounced and sparsely witnessed hops along the Shell Island beach at low tide. It reportedly rose about five feet for "some little distance," enough to show "its ability to fly." Longer flights could be attempted later. At the time, this modest effort was considered to have been the first flight of an airplane "built in North Carolina and perhaps in the South."[8]

Gouverneur, learning from his July embarrassment, was even more secretive in later experiments, for no more details reached the newspapers, though he worked on the plane for at least two more years. Chase, who soon left for other employment, later told a Wilmington historian that the project "was financially something to forget, if possible." In 1911 Gouverneur won a lawsuit for $380, with interest, against Boston's Harriman Motor Works, apparently because its motor was unreliable.[9]

In 1912 Gouverneur informed London's *Jane's All the World's Aircraft* that he was now using the *Gouverneur IV* (which implies that he had built three earlier ones), a 25-foot-long monoplane with a wingspan of 31 feet; with its pilot, it weighed 840 pounds. The latest machine, then, was 360 pounds lighter than that of 1910. In the previous three years Gouverneur had experimented "on a large scale, using electric cars for testing surfaces, *etc.*" According to Chase, his former partner hired an experienced pilot to fly one of his planes, apparently the *Gouverneur IV*, but the aviator crashed "and Gouverneur went no further."[10] Evidently his patented levers and ailerons did not provide the control he expected.

Like Chase and Gouverneur, rival Tar Heel builders developed their own, often highly idiosyncratic designs. The process usually required many modifications and stages of design before the machines were ready for trial or the builder was too frustrated to continue. Lacking money and expertise, many projects bumbled along for a few months or a year or so and then collapsed. But these home-

'spun experimenters were not cranks. All of them were amateurs at this stage, and what seemed an unwarranted notion or a useless innovation now and then proved to be well-founded intuition.

The pitfalls of trial-and-error methods are seen in the efforts of another Wilmingtonian to build his own plane. David Palmgren, a diamond buyer for a local jeweler, was a Swedish immigrant. In 1909 he made a model of his plane and persuaded some Wilmington businessmen to support his project. Founding the American Aeroplane Company, they put up a shop and hangar at Castle Hayne, near the western edge of town. They began work on the plane's structural parts and ordered two revolving cylinder, fifty-horsepower motors from the American Motor and Airplane Company in Denver, for $2,500.[11]

Framed with tubular steel, the *American*'s 30-foot fuselage rose 18 feet from the ground in front and 10 feet in the rear—with a pair of 36-foot upward-folding wings. A shaft running the length of the machine connected a forward two-bladed propeller with a rear three-bladed one, each 7½ feet in diameter with its own motor. Weighing half a ton, twice as heavy as the standard Curtiss biplane, the craft, an observer said, looked rather like "a small [horizontal] Eiffel Tower, which the wings surmount." The rationale for the radical design was that it gave the plane an extremely low center of gravity, which, with wings folded, would act "as a parachute" in case of an emergency. Its "horizontal and vertical guiders" were embodied in a foldable fish-tail rudder at the rear.[12]

On a triangular floor beneath the frame was a platform of steel tubing that carried twin copper tanks holding sixty-five gallons of gasoline and room for up to twelve standing-room-only passengers. A reporter thought that this gave the plane "the appearance of a huge gnat struggling to carry off a cage full of human victims"—an apt analogy. Four foam rubber–tired wheels with shock absorbers, the rear wheels capable of directional turning, enabled the machine, with wings and tail folded, to run on roads and streets at twenty-five miles an hour.[13]

In operation, the engines, with the pilot seated midway between them, were geared together so that, if one stopped, it acted as a flywheel while the other kept going. Stopping the front motor and slowing the rear one, while tilting the wings forward, would allow the machine to "stand still in the air." Of the many airplanes built in this fertile era, none was more unusual than the Palmgren *American*.[14]

With war brewing in Europe and the army buying its first planes from the Curtiss Aviation Company and the Wright brothers, Palmgren publicized the potential military uses of his plane. Its platform could carry a squad of armed men, and its ability to stand still in the sky made it an ideal platform from which to drop bombs.[15]

Itinerant airman Jay Reed, who visited Wilmington in the late fall of 1912, thought it unflyable and spent several weeks helping Palmgren alter it. In January 1913 Reed, at Castle Hayne, gave the *American* what was apparently its only test. He reported that, at thirty miles an

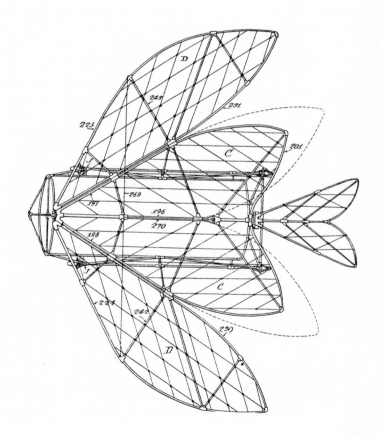

**Top view of David Palmgren's *American* patent. German Patent Archives, Munich.**

hour, its front wheels rose six feet or so off the ground but the rear wheels remained on the runway. Fearing an accident, he leaped out, and "the machine promptly sat upon its tail, wrecking it."[16]

Reed still thought that the design might work if its flat wings were cambered (slightly arched) and its height lowered by half. (His view is suspect since he had recently seen and praised the thirty-seven-winged Batson seaplane in Savannah.) Palmgren seems to have adopted Reed's ideas and begun rebuilding his machine, only to give up the project not long afterward.[17] Nonetheless, as we shall see, this invention would enjoy a moment of national celebrity.

The basic requirements of aerial stability and control were probably met by three men in their early twenties who built a plane in Kinston in the winter of 1910–11, though they claimed no originality. Ross and John Bailey and Marsh Gray said that they had obtained plans for their French Demoiselle from its designer, Brazilian aeronaut Alberto Santos-Dumont. The wealthy heir of a coffee planter, Santos-Dumont was known to provide his blueprints gratis to anyone who applied for them. But a more accessible source was the June 1910 *Popular Mechanics*, which carried the complete plans for that model.[18]

The first airplane to fly (in 1906) after the Wright machines, the Demoiselle was a small monoplane that, by 1910, was being manu-

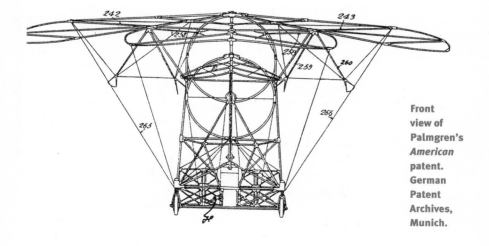

Front view of Palmgren's *American* patent. German Patent Archives, Munich.

# Working Drawings of the "Demoiselle"

### SANTOS-DUMONT'S
## Remarkable Aeroplane

**Price $2.00 Postpaid**

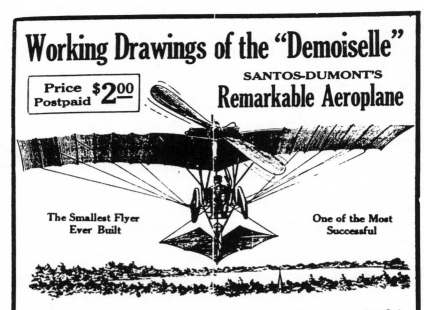

The Smallest Flyer Ever Built

One of the Most Successful

Complete plans for the construction of the wonderful monoplane offered to the public for the first time.

The machine is unencumbered by patent rights, the famous aviator preferring to place his invention at the disposal of the world in the interest of the art to which he has devoted his life. These plans are published in two parts, in the June and July numbers of Popular Mechanics Magazine, but the restrictions of space make them much smaller. They were secured by representatives of Popular Mechanics from Santos-Dumont, and are the result of consultations with his engineers and observations made at his workshops.

*The set comprises seven large blue prints, showing every detail of construction, accompanied by a description of how to build.*

## POPULAR MECHANICS COMPANY
### 225 Washington Street, CHICAGO, ILLINOIS

The Kinston airplane builders of 1910–11 probably bought the Demoiselle's plans shown in this *Popular Mechanics* advertisement of June 1910, plate 1.

factured in France and selling well, the first production airplane on the market. Although it could be a devil to fly, it introduced many audacious souls to the glories of powered flight.[19]

A photograph of Kinston's Demoiselle, one of several built by American amateurs about this time, shows it resting in the middle of Queen Street, the town's main thoroughfare, in March 1911. Built of bamboo poles and copper wire, its length and wingspan both measuring twenty-four feet, it seems to be a careful reproduction of the original design, except for its power plant. It had a secondhand fifteen-horsepower automobile motor, though the smallest used by Santos-Dumont was of seventeen horsepower and he recommended twenty-five. Twenty-two-year-old Ross Bailey, arrayed in the cockpit in a top hat with motorcycle goggles around it, appears to be a model of the confident young airman.[20]

Bailey gave the plane its initial ground test along Queen Street on March 1, 1911, after the mayor ordered the route cleared for that purpose. The Demoiselle ran nicely several times up and down the street, and its builders announced that the initial flight test would take place the next day on a field just across the Neuse River from downtown Kinston. When high winds postponed the test, Bailey made another Queen Street ground test to mollify his audience, this time breaking the homemade propeller by swerving into a telephone pole. A spare propeller was promptly mounted and the flight test rescheduled for Saturday, March 11.[21]

On the eleventh, before hundreds of Kinston residents and rural visitors, Ross drove his machine down a wooden ramp and sped across the field. But he failed in several attempts to gain enough speed to lift it off the ground. Although the builders vowed to order a larger engine, they soon concluded that the pilot's position beneath the motor, especially a heavier one, might prove fatal in an accident. The three young men abandoned the project, and the plane was later broken up because it took up too much space in the Baileys' small backyard. They lightheartedly claimed to have made "the first unsuccessful airplane flight in the history of North Carolina," though the Wrights had prior claim to that distinction.[22]

The Demoiselle on Queen Street in Kinston, March 1911. Pilot and cobuilder Ross Bailey sports a top hat and goggles. Courtesy of Esther Bean.

## BEACHEY'S BLOOMERS

Several other Tar Heel inventors or projectors of aerial machines had no better success. At Wilson in 1910, two young businessmen began building an airplane in their spare time but seem to have left it incomplete. In Charlotte in July 1912, Buick dealer Charles C. Coddington, a New Jersey native, and his associate Magnum Webb finished a biplane and equipped it with a fifty-horsepower Maxim engine but failed to get it airborne.[23]

Seventeen-year-old Hazel Worley, also of Charlotte, a youth of "remarkable mechanical genius," built a rubberband-powered model in May 1910 and coaxed it to rise about five feet for a short flight before it landed smoothly on its four wheels. Worley and his friend Ed Robbins then designed a full-sized monoplane and began constructing it in the Coddington-Webb hangar in southern Charlotte. In 1912 the Coddington-Webb interest was bought by George A. Gray, of Gastonia, who apparently provided an infusion of funds, but there is no indication that Worley and Robbins ever finished their machine.[24]

Meanwhile, another Charlotte seventeen-year-old, Johnny Crowell, began building a Blériot-type monoplane in 1911. After two years of work, he completed the frame but could not afford a motor because the war had forced up prices. When a chance came to work for an aerocompany at College Park, Maryland, an important aviation center, he suspended, then quit, his project.[25] In the 1920s and 1930s Crowell, now back in Charlotte, became one of the nation's premier stunt pilots and a gifted technological pioneer in aeronautics and aviation.

Many of these machines were directly inspired by the exhibitions of professional airmen, notably Lincoln Beachey. His show at Winston-Salem in April 1911, for example, led immediately to the founding of a company to build planes there. In May R. Duke Hay, president of the Acme Motor Company, announced plans to build and sell airplanes as well as automobiles. Soon after a November exhibition by a another professional, automobile mechanics W. H. Sloan and John Lehman formed the Winston-Salem Aerial Company. They meant to buy a Curtiss and, "if encouraged," build, sell, and exhibit Curtiss-type planes. Neither company reached its goal and apparently scrapped its plans by the end of the year.[26]

Meanwhile, James Nicholson Umstead Jr., the twenty-five-year-old son of a prominent Durham family, chartered a Curtiss after watching Beachey perform in Durham in 1911. Umstead announced that he would build and sell auto-monoplanes (for driving as well as flying) of his own design in Durham, beginning in 1912. He knew of no reason, reported the *Durham Morning Herald*, "why the wings of the machine may not be made to fold themselves up as they come to earth and drop into such a space as will not put the on-looker in needless jeopardy." It was his aim to make his machines "descend to earth when air currents are hostile" and "take the air when the roads are bad." In North Carolina, this would mean a great deal of flying.[27]

The Umstead Aviation Company hired a pilot, lured two mechanics named Teer and Conway away from a local automobile firm, bought a Curtiss, and undertook some barnstorming in late 1911 in Kentucky, Indiana, Iowa, and elsewhere in the Midwest. Probably finding that his costs exceeded his income, Umstead appears to have disposed of the plane in early 1912. He then turned to real estate and other more mundane enterprises.[28]

Such disappointment was common to aviators and promoters, but the stakes grew more tempting as speed and distance records for planes fell almost weekly. In the autumn of 1911 Cal Rogers flew a Wright biplane on a forty-nine-day ordeal of crashes and breakdowns from the East to the West Coast. Several women, including America's Harriet Quimby, were flying planes, and governments were forming air services to explore the military uses of aircraft. On March 23, 1911, Frenchman Louis Breguet took eleven passengers aloft for half a mile in his straining machine. The airplane as a tool of commerce as well as of war could not be far off.[29]

Predictably, aerial activity around Asheville was tourist-oriented. In early 1910 railroad entrepreneur R. Stanley Howland, promoting the village of Weaverville, just north of Asheville, as a resort, bought New York rights to build a plane for local exhibitions. He contracted to build a fifty-pound Witteman glider and made plans for a Farman-type biplane at Weaverville, its motor and metal parts to be shipped from New York. A hangar was constructed on a lot adjacent to the Weaverville railroad depot.[30]

The glider, the first in the state since the Wright brothers' of 1903, reached Weaverville in March. It was a small machine framed in aluminum alloy "and the lightest kind of wood," with twenty-foot wings. Howland canvassed teenage boys to try out as pilots. The coveted mantle of chief pilot fell on the shoulders of a fifteen-year-old youth named Reeves, who proved best able to handle the controls. Reeves and others made exhibition flights during March at nearby Reems Creek, a tributary of the French Broad River. The glider was briefly idled when a comely young lady sat on it, breaking three of its ribs. There was talk of installing a ten-horsepower motor for longer flights, but the plan did not materialize.[31]

Howland's biplane was made from "specially kiln-dried" local lumber and, except for its motor, was completed in June 1910. It had thirty-foot wings braced with piano wire, a forward elevator, and a balancing tail twenty feet behind the wings.[32] But no more was heard of it, perhaps because the motor arrived too late for the tourist season or not at all. At any rate, Buncombe County would have to wait for its introduction to airplane flight until Beachey's appearance a year later.

All the while Greensboro was in an airplane tizzy, owing partly to a Beachey exhibition there in mid-1911. The previous year African American inventor W. F. Johnson, a twenty-three-year-old employee of Greensboro's Judge James E. Boyd (who carried Robert E. Lee's surrender to Ulysses S. Grant at Appomattox), had conceived the idea of an electrically powered biplane. It was to be "absolutely safe from dangers of turning over in the air or falling." Johnson habitually carried with him a small book to jot down such concepts as his flying machine, a reversible streetcar switch, and a folding window screen. His airplane design had two pusher-type propellers turned by a one-horsepower electric motor. A second motor was carried in case the first overheated. His wooden model, exhibited at Greensboro's Central Carolina Fair in November 1910, won a first prize, but he apparently went no further. Johnson seems to have been the first African American to design an airplane.[33]

In the summer of 1911 teenagers Forrest E. Wysong and his younger brother Paul, under Lincoln Beachey's influence, built a glider to use on hills near Greensboro. Both boys, like the Wrights, were skilled cyclists and had recently pedaled from Greensboro to Charlotte, but Forrest, an erstwhile machinist in his father's Greensboro factory, was the principal builder. A November accident, caused by a stall, left him with a fractured skull and broken arm that halted the flights. As a senior in 1915 at the North Carolina College of Agriculture and Mechanic Arts, Forrest built a Curtiss-type biplane, Raleigh's first home-built flying machine, with plans obtained from Beachey.[34]

Automobile dealer and oilman Lindsey Hopkins, a Rockingham County native, was a dynamic entrepreneur who got his first job as a printer's devil in Greensboro in 1898. In July 1911 he formed an aviation company and ordered two Curtiss biplanes for exhibition. The

Lindsey Hopkins Aviation Company's Greensboro show in November was outstanding, with Hopkins's chief mechanic, Charlotte native Thornwell H. Andrews—the state's first native professional pilot—at the controls.[35]

Hopkins announced plans to build airplanes of the Wright, Blériot, and Curtiss types in Greensboro and Atlanta. By 1912 his company was hailed as "one of the best known in the country," having given "more successful exhibitions than any other." But investments in Coca-Cola stock and other enterprises enabled Hopkins to retire in 1913 as a thirty-four-year-old millionaire. Characteristically, he sold his remaining plane to Andrews at the point when enthusiasm for exhibitions began tapering off. He later served as a director for North American Aviation and the Sperry Corporation.[36]

North Carolina journalists generally embraced airplanes as an important step forward, though not all of them were convinced that flying could be made practical. As late as 1910 the *Asheville Citizen*, a firm holdout, contended that "this generation will not see the perfection of the flying machine." The editor believed that airplanes would never "come up to practical use for commercial purposes."[37]

The *Winston-Salem Journal* was among several papers that chortled over the *Citizen*'s stance. The *Journal* replied that future planes would take off from water and that "air lines for the carrying of passengers will one day become a reality." To the *Citizen*'s view that, if God wished men to fly, He would have given them wings, an editor shot back that the want of "steam-heated legs" did not prevent men from racing along rails at sixty miles an hour. The *Wilson Times* praised airplane safety, a virtue founded on the cheerful belief that, in an accident, "the air resistance will prevent a too rapid descent and injury to the occupant."[38]

Within two months after the *Citizen*'s latest contrarian sallies, Charles K. Hamilton's round-trip flight between New York and Philadelphia in June 1910 left the Asheville editor eating spruce-and-cotton crow. The *Charlotte Observer* and *Raleigh News and Observer* were among leading state papers that promoted heavier-than-air flight.

Two North Carolinians who gained aeronautical notoriety in these groping years, James Spencer Spainhour and William W. Christmas, developed advances in control that eluded fellow Tar Heels and many others. Both men early moved north of the Mason-Dixon line and each, probably for that reason, enjoyed greater success than rivals back home.

Spainhour's father, well-known Iredell County ethnologist Dr. James M. Spainhour, was a friend of Samuel Langley, whose work aroused the son's interest. James S. Spainhour was a Lenoir native and a product of the North Carolina College of Agriculture and Mechanic Arts. While working in 1911 as a Westinghouse engineer in Pittsburgh, Pennsylvania, he designed a monoplane reminiscent of Blériot's cross-channel type XI machine. He was especially interested in stability in flight and, in 1914, patented a device for that purpose.[39]

Spainhour built his plane on a farm outside Pittsburgh in the summer of 1911. He equipped it with a device by which, when air currents forced a wing upward, compressed air could "warp the other wing in the opposite angle, thus preserving perfect balance." In addition, foot levers could also be used to aid wing warping. Power was provided by a fifty-six-horsepower engine supplied by John Kowalsky, another airplane builder in the Quaker State.[40]

The plane was finished in late summer, and Spainhour made several test hops on September 21. On the last of these, he took off on what was intended to be a short flight at an altitude of about thirty feet. But the plane behaved so well and the weather was so ideal, that he was emboldened to fly higher. Corkscrewing upward over the two-acre field at about fifty miles an hour, he reached an altitude of one thousand feet before starting down. While descending, according to a Pittsburgh paper, he "dipped and circled and played about in the air" before landing at a point half a mile from his takeoff. It was a remarkable flight for a pilot who had never flown before.[41]

Spainhour's "automatic stability device," predicted one of his backers, "will revolutionize the entire aviation game." North Carolina's *Morganton Herald* anticipated that he was likely to "make much

James S. Spainhour (standing) with his airplane, 1911. A Lenoir native,
Spainhour studied engineering at present-day North Carolina State University.
Courtesy of Mrs. Gloria L. Sodergren.

money out of it." In fact, however, most airplane builders were already using ailerons on the trailing edges of their wings and achieving better results with them than the Wrights' cumbersome "wing-
warping" system.[42]

Spainhour moved on to Hempstead, Long Island, which had re-

Fannie Spainhour beside her husband's monoplane near Pittsburgh, Pennsylvania, September 1911. Courtesy of Mrs. Gloria L. Sodergren.

cently become America's major aeronautical center. In April 1912 he was said to have made a flight of half a mile, but a subsequent crash seems to have put an end to his project. In June 1918 he received a patent for interchangeable rudders and elevators, eliminating the need for pilots to carry "spare parts such as a rudder of one design and elevators of another design." During World War II, he worked with the supersecret Manhattan Project that in 1945 produced the first atomic bombs. He died on Long Island in 1946.[43]

William Christmas, who knew Spainhour, had purportedly been studying bird flight since he was a teenager. He was said to have built a glider in 1905 and then, emboldened by his own and the Wright brothers' successes, began building a powered plane. He was understood to have been an associate of Samuel Langley in the Potomac River experiments, and to have from him a "bequest of aeronautical data," though he had become disillusioned with Langley's "tandem monoplane" design. The result was his *Red Bird I*, of 1908, a biplane similar to that of the Wright and other machines of the time.[44]

Christmas claimed two unique features for his own plane. Its deeply cambered wings were of cantilever construction, meaning that the wing supports were contained within the wing. They also carried inset ailerons, hinged sections inserted into the trailing edge of each wing to control the plane's roll. The first of these features gave his wings more strength and lift capacity, while dispensing with wires and other exterior supports. His ailerons, for the first time inset into the wings rather than simply attached, represented a decided improvement over wing warping.[45]

Seeking privacy in his work, Christmas is said to have arranged with a farmer, Robert Nevins Ions of Fairfax Station, Virginia, near

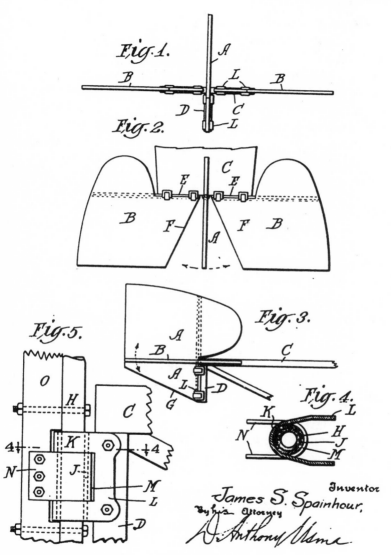

Washington, D.C., to construct the machine on his woodland farm. *Red Bird I* had twin pusher-propellers driven by chain transmission from a homemade two-cylinder, fifteen-horsepower motor mounted behind the wings.[46]

After ground tests, the builder, allegedly on March 8, 1908 (or perhaps several days earlier), made a low flight of several hundred feet, with a forty-five-degree turn. "Scared?" he later commented. "I had too much to think about. All I wanted to know was whether it would get off the ground or not." The flight reportedly was witnessed by Ions and his wife and two farmhands. Christmas made several more hops in the next few days, one for half a mile and up to twenty feet in height, before "she got away from me. One wing hit a tree. I was . . . going about thirty-five miles an hour. Of course the plane smashed, and I slithered out along the ground." He was said to have burned the wreck to protect his ideas and started working on an improved version.[47]

By virtue of this claim, Christmas boasted that he was the first American after the Wrights to build and fly his own plane, a feat he accomplished four days before the first flight of Glenn Curtiss on March 12, 1908. Except for a few like Arch Whitehouse, who writes that "there can be no doubt" about it, the claim is almost universally dismissed by air historians, even though it has received some independent corroboration.[48]

When Robert Ions died in Richmond, Virginia, in 1949, his eulogist stated that Ions and Christmas built a model plane at the Ions farm in mid-1908 and a full-scale biplane later that year. Its only flight was reportedly made—by Ions—in late 1908 at Anacostia. If so, then the Christmas-Ions machine flew well after Curtiss, F. W. Baldwin, and others in that year and was perhaps the sixth or seventh American one to fly in 1908.[49]

In 1909 Christmas publicly stepped into the field of aviation, proclaiming that he had spent over a quarter century studying the subject. The army in that year established an aerial training facility at College Park, Maryland, and allowed civilian airmen to use the field. Christmas thereupon founded the Christmas Aeroplane Company, with offices in Washington, D.C., and New York, and began constructing an improved biplane at College Park, the *Red Bird II*.

Christmas claimed that this plane, like Palmgren's, could stand still in the air.[50]

On October 15, 1911, a mechanic, intending only to taxi the new plane around the field, inadvertently (he said) took off, flew it for a mile on its maiden flight, and brought it back undamaged. "Langley's words" of encouragement, exclaimed Christmas, "have all come back to me and I feel that I have reached the goal which I started after fourteen years ago."[51] (He does not sound like someone who has enjoyed prior success.)

In 1910 Christmas applied for a patent, issued in 1914, for his hinged aileron, a method of control that soon became standard on virtually all airplanes. Arch Whitehouse wrote that the 1908 *Red Bird I* was "unquestionably the first American aeroplane to be fitted with ailerons for lateral control that were in a hinged portion of the wingtips, just as they are employed today." He also believed that this machine "had elevators mounted as they are now as an integral part of the tail assembly, as they are on conventional planes." Christmas may not have been the first to employ this method of control, but he was the first to patent it. Reportedly he later sued the federal government for unauthorized use of his invention and won a $100,000 settlement, but the allegation is unverified.[52]

## THE GRAND CENTRAL PALACE EXHIBITION

*Red Bird III*, finished in early 1912, was the first of the Christmas planes considered suitable for carrying heavy loads on long-distance flights. It was the most credible and widely publicized, so far, of his machines, despite his insistence that it would circle softly to a landing even if the motor stopped "in midair, or the operator faints away." It was displayed at Washington's first aero show in late 1911. When New York City announced that the first international airplane show would be held at its Grand Central Palace exhibition hall in early May 1912, Christmas shipped his new plane there in the hope of attracting financial support for further ambitious enterprises.[53]

On the eve of the New York show, the U.S. Post Office signed a contract with Christmas for *Red Bird III* to carry the first cross-

country airmail ever attempted. When the exhibition closed, the plane was to fly from the Mineola airfield on Long Island to Pennsylvania Avenue in Washington, D.C., with one hundred pounds of mail (10,000 letters), alighting triumphantly at the very steps of the Post Office Department. It was not only an exciting experiment for the future of communication but also a public relations bonanza for the fledgling Christmas Aeroplane Company. A successful flight would entitle Christmas to establish regular airmail service from Washington to Norfolk and Baltimore.[54]

The Grand Central Palace exhibition provided the largest array of flying machines and equipment ever yet assembled in America under one roof. It had toy planes, passenger balloons, dirigibles, Tom Baldwin's *Red Devil* biplane, the Blériot monoplane in which Harriet Quimby had recently become the first woman to fly across the English Channel, parts of the 1903 *Wright Flyer*, motors, propellers, wheels, and more. Twenty-three foreign and American airplanes were among the fifty-five exhibits.[55]

The ten-day show, overshadowed by the April 16 sinking of the steamship *Titanic* and rescue efforts, drew smaller crowds than expected. But it was something of a triumph for Christmas. Visitors found the *Red Bird III* a big machine, said by its builder to be the largest in the world. He called it "the only self-balancing machine in existence," one that "can't be turned over in flight." It would carry three people and had swept-back wings and a fifty-horsepower ro-

1,095,548.

Patented May 5, 1914.

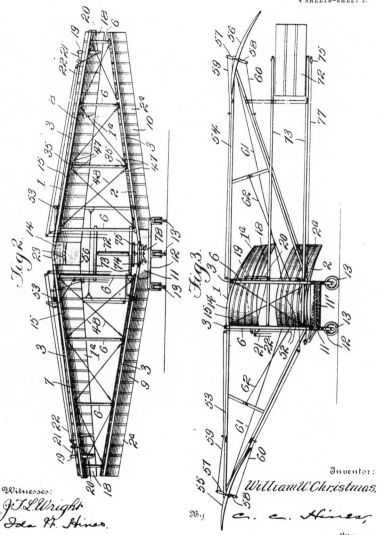

Witnesses:
J.T.L.Wright
Ida M. Hines.

Inventor:
William W. Christmas.

By
C. C. Hines,

In 1914 Christmas was issued the first patent for the hinged aileron, a feature on virtually all airplanes since then. U.S. Patent Office, Washington, D.C.

tary engine. Its ailerons were said to be revolutionary, automatically adjusting to fluctuating air pressure.[56]

An added attraction at the Christmas exhibit was the presence of telegrapher William Russel, who demonstrated how, as a passenger on the Washington flight, he would stay in touch with both New York and the capital by means of wireless messages. Paul Peck, a reputable Washington, D.C., pilot, was signed by Christmas to fly the plane, which would follow the route of the Baltimore and Ohio Railroad to its destination.[57]

The mail flight, initially set for May 23, began to encounter delays. Christmas, concluding that the flight demanded a larger motor, had the 50-horsepower engine replaced with a more costly 100-horsepower mechanism. Then Peck either resigned (perhaps because he had not been paid) or was fired; he was replaced by Clinton O.

Christmas in front of the *Red Bird III* just before it was exhibited at New York's Grand Central Palace in May 1912. From *New York World*, undated tear sheet in Orren Randolph Smith Scrapbooks, Military Collection, Civil War, "Historical Odds and Ends," box 5, folder 2, p. 102, North Carolina Division of Archives and History, Raleigh.

Hadley, who needed time to become familiar with the plane before attempting a long-distance run. The flight was finally scheduled for the first week in June.[58]

Hadley took off alone (telegrapher Russel evidently opted out) early on June 6 for Washington but, quixotically, decided the weather was so fine that he would attempt a world-record endurance flight before leaving, presumably an added, though irresponsible, public relations fillip. He reportedly stayed up for 185 miles; his time in the air—four hours, twenty-nine minutes, and thirty seconds by his account—was only half a minute short of "the American record at that time."[59]

As he approached the record over Long Island, Hadley heard a clink, a crack, and another clink, and his engine went dead. He was able to glide from an altitude of seven hundred feet to a safe landing, but examination was said to have revealed a spent bullet in a cylinder and a broken propeller. This indicated that someone, presumably a rival, had sabotaged the flight. Christmas offered a $1,000 reward for information leading to an arrest, but the Post Office apparently concluded that it was wasting time on him and canceled its contract with the inventor.[60]

The result was not only a crippling blow to the Christmas Aeroplane Company but also a source of further suspicion directed at its president. The firm was out of money, and by now it was common knowledge in aviation circles that Christmas's claims of enormous advances in aeronautical science were not corroborated in public performances.[61]

Christmas professed to be a graduate of George Washington University and the medical school at Johns Hopkins. Inquiry would have shown that Johns Hopkins had no record of his attendance but that he graduated from George Washington in 1905 at age forty. There seems to be no record of his alleged association with Langley nor independent testimony to his years of scientific bird-watching and aerial experimentation before 1908. His self-balancing *Red Bird II* was erroneously said to become steadier the harder the winds blew and to "remain perfectly still in the air when the engine is inoperative." Christmas boasted that *Red Bird III* could be enlarged to lift twenty tons "or to infinite dimensions." He boldly asserted that the aviation industry was virtually "founded on [his] inventions and discoveries."[62]

"A huge man with sloping shoulders slouched forward so that his hands seemed to fall almost to his knees," Christmas could be physically and psychologically intimidating. Wealthy businessmen and bankers who were respected for their entrepreneurial acumen readily yielded to his grandiose schemes for inventions. He "looked at you with a heavy, hypnotic stare," wrote a former associate, his "domineering manner" abetted by "a heavy abrasive voice."[63]

Christmas was a man of considerable technological talent and by no means simply a fraud, but he had an incorrigible penchant for referring to his modest achievements and, too often, Rube Goldberg ideas, as earth-and-sky-shaking discoveries. Over the coming years, his reputation sank as his claims grew more fantastic. Only a few of his three hundred alleged "inventions" were ever awarded patents, and none appear to have been commercially successful.[64]

This judgment of history, however, may not be altogether fair. One of his main goals was to build low-speed airplanes to carry heavy loads with a minimum of fuel, as do large birds. He asserted, for example, that *Red Bird III* could stay airborne at twenty-seven miles an hour.[65] Such a plane, if successful, might have helped pave the way for passenger and cargo planes that were able to compete more effectively with railroads and trucks. But the trend in aeronautics, which was heavily influenced by the experience of World War I, was toward speed and agility. In any event, Christmas would soon reappear with designs and claims that were far more spectacular than those he had already made.

Christmas was not the only Tar Heel to show an airplane at the Grand Central Palace exhibition. Whereas his plane was the largest in the show, David Palmgren's was by far the tallest. The Wilmingtonian transported his monoplane, the *American*, by steamboat to New York, though it had not yet flown, in the hope of finding backers to make it fly. His plane attracted nearly as much attention as Christmas's, chiefly because of its unorthodox construction. It was said that the *American* would be taken to Governor's Island after the exhibition and "given a trial flight."[66]

A New York wit, noting that its height and tubular steel frame

Wilmington's David Palmgren displayed his unorthodox monoplane, the *American*, at the Grand Central Palace exhibition in New York in May 1912. From *Scientific American*, May 25, 1912, 472, Library of Congress, Washington, D.C.

made it look rather like the trucks used to maintain New York's power lines and elevated railways, proposed that the *American* be used for such purposes if it failed to fly. The *New York World*, tongue in cheek, said that Palmgren's motors "will never stop as it is impossible for [them] to be mended in mid-air. . . . So they must not stop." Despite a respectful writeup and photograph of his plane in the *Scientific American*, Palmgren went home disappointed.[67]

Like Christmas, Palmgren had a shadowy past. He claimed to have earned bachelor's and master's degrees in mathematics from Sweden's prestigious University of Uppsala and to be widely known in European scientific circles through his scholarly articles. But the university has no record of his attendance, and Swedish aerohistorians know only of his work in Wilmington. Later claims by his son that Palmgren's plane flew on several occasions are patently false.[68]

Palmgren hoped to win a $15,000 prize that millionaire Jay Gould offered for "the most practical heavier-than-air machine to be equipped with two or more motors working independently." But Palmgren's claim that his plane would not turn over in flight depended too much on the fact that it could not fly. In 1917 a group of Palmgren's stockholders proposed to donate his "unfinished machine" to the federal government, but Washington showed no interest.[69]

North Carolina made an impressive appearance at the Grand Central Palace exhibition, but it marked the end of the state's most prolific era of aeronautical design and construction before the late twentieth century. From 1910 to 1912 North Carolinians produced only two successful airplanes (Spainhour's monoplane and Christmas's *Red Bird III*), and the growing sophistication of flying machines was rapidly leaving Tar Heel capabilities behind. No North Carolinian, with the exception of Christmas, had an industrial base or financial resources to match, let alone anticipate, advances realized in aerial technology during World War I and afterward. As elsewhere, by 1912 the state's larger towns were growing a bit jaded by airplane exhibitions, actual or attempted. Few planes did more than take off, make a turn or two, and land. And many airmen found that exhibition flying was an exhausting business and extremely dangerous.

Fliers charged the sponsors of exhibitions substantial fees for their appearances. But because tickets could only be sold to spectators inside exhibition fences, sponsoring groups often lost money. Rebirth of the spectacle would await the barnstorming era after World War I, when safer, more agile planes and better fields and facilities were available. For the time being, aeronautical progress depended mainly on Europeans, their incentive sharpened by the approach of war, their governments providing strong support.

The most innovative and prospectively important advances of the early North Carolina airplane builders appear to have been the all-metal machines of Gouverneur and Palmgren; innovative approaches to stability and control by Gouverneur, Spainhour, and Christmas; and the Christmas patent for the inset aileron. It has not been established that any of the four men was first to explore or utilize these techniques, nor can it presently be shown that their efforts influenced others. If not first, however, they performed their best work, like Tar Heel helicopter designers, at the leading edge of aeronautical technology.

Owing to the character of journalism in that era, the North Carolina public knew virtually nothing of the scope and variety of aeronautical work across the state. Small newspapers could afford to exchange issues with only a few others. For instance, Greensboro knew very little of Charlotte's aerial enterprises, and neither had more than an inkling of those in Asheville or Kinston, let alone Pittsburgh, Pennsylvania, or Washington, D.C. Even the experimenters themselves were largely unaware of one another. As a result, North Carolina's vigorous aeronautical activity of this epoch remains absent from state histories and barely noted, if at all, in the annals of a few municipalities and counties.

Much the same situation persisted throughout World War I. But the wartime call to patriotism demanded a wide sharing of tales of soldierly derring-do, and no such stories were more avidly consumed than those of people who flew airplanes. By the war's end, moreover, it could be shown that Tar Heel fliers ranked among the world's best, one of them, in fact, arguably the finest of all. In this sense, North Carolina would remain for the time being a significant presence in the field of aviation.

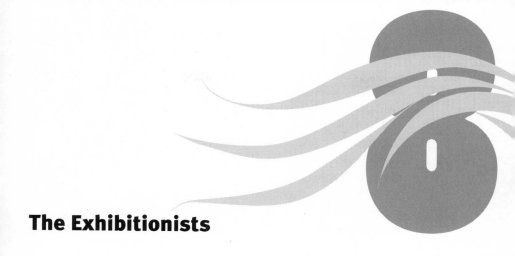

# The Exhibitionists

### HEAVEN KNOWS, MR. ALLISON

Raymond Vance Allison was a young man of whom western North Carolina might well be proud. The son of a prominent family in Statesville, where he was born in 1886, he graduated from Cornell University in 1907 with a degree in mechanical engineering. His education placed him among the elite of a New Age technocracy consciously destined to mold the twentieth century into an epoch of amazing machines that would transform life for all humankind. Ambitious and self-assured, handsome in his Van Dyck beard, he epitomized the best of the new century's young manhood.[1]

Allison's heart was initially set on playing a role in the development of the automobile, but he accepted a position as North Carolina representative for the new K.R.I.T., a Detroit-built automobile. He returned to Statesville in early 1910, his engineering career on hold, to establish his first dealership. Beginning aggressively, he toured North Carolina by car to demonstrate the machine's virtues and became well known in and around Statesville as an enthusiastic auto racer.[2]

By mid-1910, however, not automobiles but airplanes seemed likely to become the century's most exciting technology, and Allison was soon a devotee of flight. His goal was to learn to fly and then be the first pilot to give heavier-than-air exhibitions in North Carolina. What should have been the state's first such exhibition, by the Curtiss Aviation Company at Charlotte in the spring of 1910, was canceled when the pilot backed out. But Allison had little money and would have to start from scratch. Though it was a daunting challenge, the self-confident Cornell man felt equal to it.[3]

The challenge quickly became urgent. In September 1910 Allison was stunned by news that a Charlotte paper was making a new effort to give North Carolina its first air show. As the site of Dr. Daniel Asbury's experiments in the early 1880s and the home of the *Charlotte Observer*, the first American newspaper to champion heavier-than-air flight, Charlotte deserved the honor. A Curtiss biplane would perform there on November 10. Allison, his competitive instincts aroused, had no intention of being denied his dream.[4]

Leaving almost immediately for New York, he had scarcely eight weeks to accomplish his miracle. He revealed to friends that he intended to buy an airplane, take flying lessons, bring the machine to Charlotte, and put on his own air show before the Curtiss performed. The plane would be a $7,500 state-of-the-art French monoplane, evidently a Demoiselle or a Blériot; he would return with it after a period of flying lessons. Allison appeared to have little chance of carrying out such an extravagant plan, but, as the *Charlotte Observer* remarked, he was a man "of courage and will power and has recently completed cross-country automobile tours." The paper's editor, Frank M. Caldwell, was a close friend and perhaps a source of the young man's new dedication to aviation.[5]

Charlotte was soon on tiptoe about the November 10 Curtiss show. The town, the *Charlotte News* reported on October 24, was "agog over the . . . great exhibition." Thousands would attend; baseball's Ty Cobb had been invited, and there would be lunch stands and peanut and popcorn vendors. The show, it was speculated, might even persuade Glenn Curtiss, who was planning to establish an air training school in the South, to locate his facility in Charlotte.[6]

Nothing more was heard of Raymond Allison until the morning of November 7, when he arrived in Charlotte with only two days remaining to present his air show. He informed flabbergasted journalists that he had acquired a plane, learned to fly it, and shipped it by rail. It would reach Charlotte the next day, and he would give his exhibition on the afternoon of the ninth, a day ahead of the Curtiss show. He had confounded the skeptics.[7]

But not everyone was thrilled. The *Charlotte News*, sponsor of the November 10 event, was livid over this development. At stake was not just the paper's credibility but hundreds of tickets already sold or seeking buyers, and funds invested in advertising, hiring the fairgrounds, a band, and other arrangements. The *News* angrily accused the *Observer* of putting Allison up to the stunt to harm its main competitor.[8]

Allison had an exciting story to tell reporters. His plane, he disclosed, was a Curtiss type that he had acquired from Henry C. Cooke of Garden City, New York, for $5,000—$2,500 less than he would have paid for the French plane he first intended to buy. Moreover, he was the proud owner of Lloyds of London's $5,000 "Aviation Policy #1," evidently the first such insurance plan written by the firm. He and Cooke had "stood over the workmen as [the plane] was built."

Raymond V. Allison at the wheel of one of his cars, ca. 1910. Courtesy of Thomas A. Allison.

He had had the opportunity to "drive" it only a few times, and there had been difficulty shipping it south during an expressmen's strike, which had forced him to hire a team of horses to transport it to the railroad station. "Every hope and ambition of my life has been centered around this feat," he declared. "I am not going to let anything balk or thwart me."[9]

Then came the thunderclap. Henry Cooke arrived with the airplane on the evening of November 8, but, Allison announced, two "bows" were broken in transit and could not be repaired in time for his show. Though he could not fly on the ninth, he would give a series of exhibitions at various places, "probably including Charlotte at a later date." The embarrassed and crestfallen *Observer* noted that Allison, son of former U.S. marshal Thomas J. Allison, might be consoled that "his position as the owner of a heavier-than-air plane is unique in North Carolina."[10] (By this time Weaverville's Stanley Howland may have disposed of his own plane.)

It was only the beginning of Allison's troubles. His whole audacious scheme collapsed on November 9, when Cooke, the plane's builder and owner, made a number of startling revelations. He told the *Charlotte News* that Allison had promised to pay him $1,500 of the plane's $2,500 (not $5,000) price on arrival but now refused. There had been none of the delays Allison spoke of in bringing the plane to Charlotte; it was undamaged but would require several days to assemble. Cooke had made a few hops in it but no genuine flight and would be unable to give a show any time soon. Finally, Allison had never flown this or any other airplane.[11]

The *News* charged the *Observer*'s Caldwell with having tried unsuccessfully, on November 7, to rent the fairgrounds for Allison. "Since when," the *News* asked, "has a reputable newspaper stooped to rob a rival of deserved honor, or to belittle, by questionable methods, a legitimate undertaking?" Both the *Charlotte Chronicle* and *Observer* professed shocked innocence.[12]

Allison left Charlotte without defending himself, a sure sign that Cooke had not lied. Evidently Allison had expected to bring off his coup without money, gambling that he could pay Cooke from the exhibition's proceeds. He then hoped, briefly, that Cooke might agree to a new scheme to give shows in other towns, but the builder refused to have anything more to do with him. Broke, without funds even to pay express charges to return himself and his plane to New York, Cooke left for Spartanburg, South Carolina, where his craft was the first airplane seen in the state. Newspapers noted his failure to get off the ground there, and he was not heard from again. Allison returned to Detroit to help develop a new kind of truck, the Gibraltar.[13]

The Curtiss exhibition was a flop. Winds on November 10 were too strong for the pilot, J. A. D. McCurdy, to take off. The next day was worse, though he made a brief hop during a lull. Determined to stay until he flew, he tried again on November 12, making only a ninety-second hop. Skidding to a halt, he "broke a section" and declared that he would not try again. It was the first such accident in his career. Ty Cobb, needless to say, did not show up. Yet, however briefly, many Tar Heels had now witnessed a public trial of an airplane and could testify to at least the latent possibilities of such machines.[14]

November 10 happened to be the day that the Chase-Gouverner multiplane made its hops on the beach at Shell Island, east of Wilmington. But the saga of Raymond Vance Allison did not end with the debacle in Charlotte. He would soon reappear in the headlines to demonstrate both his mechanical genius and his pronounced eccentricities.

Allison's 1910 encounter with aviation illustrates the explosive transition of flight, in keeping with the new century's evolution from a heroic to a commercial age. The first Tar Heel harbinger of the new era, he had hoped to make his mark as a showman, as scores of others were doing, rather than as a builder or an experimenter. Less than seven years had passed since the first successful flight of an airplane. In contrast, it was a quarter century after the first successful gasoline automobile, built by J. J. Lenoir of Belgium in 1862, before anyone (in this case, the German Carl Benz) made money selling or driving cars.[15]

Almost beating out Wilmington's Harold M. Chase for the glory of becoming the first Tar Heel to fly with wings was Charles W. James of mountainous Madison County. The high-spirited twenty-three-year-old lived about five miles from Marshall, the county seat. In mid-November 1910 several of his brothers, sisters, and friends were on hand when he tried to fly.[16]

"Young James," says a newspaper report, "had been drinking [as had his company], when the subject of 'airships' came up, and after several attempted to explain the workings of airships, the young fellow exclaimed . . . that he would show them all how to 'fly,' and . . . he rushed upstairs." He returned with "a pair of old eagle wings from the attic," and an old umbrella; he called all to follow him, and he would "make a flight equal to any that had ever been made."[17]

James led his unsteady entourage for about a mile to a forty-foot cliff. "Here he took from his pockets several feet of small . . . rope, and fastened the eagle wings on his back, up near his shoulders, and raising his umbrella, made the 'flight.'" His leap from the cliff top began badly: the umbrella turning inside out, and "he was hurled to the bottom." The onlookers, rushing to his apparently lifeless body, found him bruised and unconscious but otherwise uninjured. Regaining consciousness, he explained that he "forgot to 'flop'" his eagle-wings.[18] Kept alive in oral tradition, James inspired poet John Foster West of Boone, many years later, to immortalize "Will Sharp" and so record James's legacy in a poem entitled "Insecure Icarus." Will, bundles of fodder tied to his arms, "got set to jump and join the birds / A / crowd had come to see Will kill himself. . . . Will crouched and leaped, then fell all a-tumble, fodder blades like fall leaves filling the air. . . . Will groaned, 'everything went right but jest one thing: I clean forgot to flop my fodder wings' / Ever after that when Will Sharp happened by / folks would flop their arms and laugh like crazy."[19]

North Carolina's first successful air show after the Charlotte and Madison County debacles was sponsored by the *Raleigh News and Observer* and opened just six days after the Charlotte rout. Some of the early air shows were staged by lone craftsmen who built their own machines. These exhibitions, first held in 1909 in northern and western cities, were often well attended and lucrative. Soon the Wright brothers, Glenn Curtiss, and other entrepreneurs were sending out teams of professional fliers and mechanics to give shows throughout the country. So far, few had appeared south of the Potomac River.[20]

The *News and Observer* was entitled to trumpet its November 16 and 17 show at the state fairgrounds in Raleigh as one of the premier North Carolina events of the new century. The team en route from Curtiss included airmen J. A. D. McCurdy and Eugene Ely. The event gained stature when, two days before his arrival, Ely executed a flight from the deck of the USS *Birmingham* in Hampton Roads, Virginia. The first flight from a ship's deck, it heightened Raleigh's awareness of what it would witness, besides setting the stage for a revolution in the military use of airplanes.[21]

As Raleigh's show approached, the *News and Observer* squeezed it for all the publicity the paper could get. Special trains were arranged for spectators from throughout the state's central region. "Raleigh Aviation Meet Takes State by Storm," crowed the paper; it would be seen by "the greatest gathering of North Carolinians ever." Persons wishing to fly as passengers were urged to write letters to the editor indicating why and be willing to sign forms releasing the Curtiss Aviation Company from claims for death or injury. Their letters were turned over to Curtiss, which would choose those permitted to fly with McCurdy or Ely. Most readers were not tempted. One sent in a poem that read, in part:

Suppose I should get in one of these things
With rudders and motors and curious wings
And when we'd gone up say fifty feet
The crazy gas engine would start to heat
Or something got out of gear just a bit
'Twould be all right when you're flying
But hell when you hit.

A disconsolate housewife volunteered her husband for a flight.[22]

On the day of the show, local schools, factories, and stores closed, and booths were set up at the fairgrounds to sell barbecue, bruns-

wick stew, and other snacks. Outlying towns were papered with bill-
boards and four-color posters. A race was arranged among seven
automobiles and another between a car and McCurdy's plane. The
latter stunt was called off when the pilot could not get airborne for
the contest.[23]

During the exhibition McCurdy and Ely each made two flights,
and the meet was pronounced a whopping success, though no passengers flew because of unsettled weather. Things were less well
managed on the ground. During the auto race, Mrs. Phil Woollcott,
wife of a Raleigh car dealer, swerved her Hudson off the track and
into the crowd, badly injuring Alice Castleberry and Ida Bryant, two
African American women from Durham, the latter fatally. This left a
bitter aftertaste, but thousands had seen their first airplanes and
could now imagine that aviation might one day play a role in their
lives. It was a letdown that two North Carolina builders, probably
Minor F. H. Gouverneur and David Palmgren, invited to show their
planes at the meet did not appear.[24]

The Raleigh air show was the first in a series of exhibitions across
North Carolina in the next three years. J. C. Storris, a New Yorker,
brought a Blériot monoplane to Goldsboro at the end of November
1910 for a four-day show. Repeated efforts got him only a few feet off
the ground, and by December 2 his machine was "almost a total
wreck." His failure was galling to the Goldsboro Argus and other sponsors of the show.[25] But then came Beachey.

Lincoln Beachey, then on his first airplane tour, would soon be
the outstanding aerial showman of the era. He had appeared at the
state fair in October 1908 with a dirigible before turning to planes.
During 1911 he would criss-cross North Carolina in shows that
brought the reality of aviation home to an excited public. The new
man at Curtiss, who had made his first flight in December 1910, he
had made dozens since. He was already earning a name for himself
as a daring and resourceful artist of the clouds.[26]

Beachey's next North Carolina show was at Wilmington in March
1911. Curtiss hired helicopter designer Frank Herbst as its regional
agent to arrange air shows. Aviation-minded Wilmington, with its
multiple aircraft projects, was advertising itself as an agreeable winter camp for northern fliers. Flights there by Beachey and McCurdy

thrilled a throng of five thousand spectators but failed financially since most watched from outside the appointed grounds.[27]

Beachey arrived in Pinehurst on March 16 and found enough interest among vacationing Yankee nabobs to open a four-week aviation school. His first passengers, strapped to the lower wing since the only seat was the pilot's, were his wife and Mrs. R. B. Middleton of New York. Taking off from golf courses, Beachey "bombed" onlookers with oranges and twice took up Commander Saito of the Japanese navy (who perhaps already dreamed of an aerial visit to Honolulu). Local photographer Ellsworth C. Eddy took the first aerial photographs of North Carolina from Beachey's plane.[28]

Golfer Charles Evans Jr., after flying with Beachey at Pinehurst, wrote that he imagined he might be stricken by the "shuddering, awesome sort of fear which assails a boy when passing a cemetery at night." In the air, however, he found himself "hovering gently . . . above the pine woods." A "profound sensation of security came over me, and while the rush of air was tremendous and the revolutions of the high-power engine deafening, I felt as safe as if sitting upon a sheltered balcony." Below, "[t]he village . . . was an exquisite bas-relief in monotone and sparkling pin points, the golf links, a beautiful stretch of green, and the roads, silver ribbons, with surrounding landscape stretching on and on like the ocean in infinity."[29]

More often, such romantic views were dispelled by a few minutes aloft, an experience both soul-stirring and terrifying. Fliers contended with a burning sun, high winds, and soaking storms. Skies that looked calm from below might actually be minefields of invisible turbulence that could generate unexpected dives and stalls. Flimsy wings could collapse, controls jam, wires snap. Because North Carolina had no airfields, pilots trying to land—whether in an emergency or not—risked hitting trees, stumps, unseen ditches, or other hazards. Jumping out was not a solution since there were no parachutes. "Everybody," said a professional aviator, expected to be killed and many were—529 in 1911 alone.[30]

Beachey was in Greensboro on April 7, Fayetteville the next day (where, "owing to differences with the management" of the fair, he refused to fly), and then Asheville, where he was the first aviator to soar above the Blue Ridge. He flew in Durham on May 3 and, after a

Lincoln Beachey prepares to take Commander Saito of the Japanese navy on a demonstration flight at Pinehurst in March 1911. Courtesy of Mrs. Ellenore Eddy Smith.

summer's absence barnstorming in northeastern states, in Salisbury on September 21 and 22, soaring three thousand feet above the town and performing deft dips and circles for eight thousand people below. The *Salisbury Evening Post* called it "one of the greatest sights ever seen in Salisbury."[31]

The Durham show produced tremendous excitement. *Durham Sun* reporter Tom Bost estimated the crowd at East Durham's baseball park to number eight thousand, about a quarter of whom paid to get inside the fence. Beachey gave a brief talk on flying, "enough to keep one from calling him an oyster, and little enough to keep anybody from calling him a fool." Though a modest and diffident man, he "disports a diamond almost as big as his gasoline engine." Helpers twirled the propellers and held the plane down until Beachey was ready. He took off from near the third base line and in moments was out of sight from the grandstand. In half a dozen circuits of the field, he covered about twelve miles in as many minutes. At one point, irked by the sight of freeloaders on rooftops, he dropped from eight hundred feet in a steep dive, swooping "like a

hungry hawk reaching for an unripe gosling," and cleared a barn roof of spectators. As he landed, an aide caught a wing to stop him but was jerked down and run over, though not hurt. Bost thought it was the first airplane flight from a ballpark.[32]

Already, however, some people felt that the air show was not fulfilling public expectations. Winds and breakdowns postponed or canceled many flights and shortened others, upsetting patrons, many of them from considerable distances. A Greensboro editor was miffed when Beachey made two flights instead of a contracted three and stayed in the air only thirty minutes instead of two hours; those arriving late missed even that. The editor groused: "There is a growing impression . . . that the flying machine business is a graft, pure and simple. The flying machine may be all right but it seems to be promoted by grafters."[33] It was a cheap shot at responsible airmen.

It was a lot to ask a pilot, sitting alone on the edge of a wing of spruce sticks and cotton, to do death-defying stunts. But enterprising fliers scoured their imaginations for tricks, often with highly gratifying results. In August 1913 DeLloyd Thompson executed

North Carolina's first loop-the-loops at the state fairgrounds. A Barnum and Bailey balloonist sent a horse aloft at Greensboro, and balloon races were held in various towns.[34]

At New Bern in November 1911, exhibitionist Eugene Heth took up local photographer Bayard Wootten, apparently the first North Carolina woman to fly in a plane and take aerial photographs. "As the big machine soared over the race track far into the clouds," wrote a reporter, "the spectators craned their necks and followed its flight with their eyes until it [descended] and Mrs. Wootten alighted amidst the cheers from many throats."[35]

Itinerant airmen also induced a mild recurrence of UFO sightings. A phantom biplane flew over Greensboro's Pomona suburbs at sixty miles an hour on October 9, 1911, heading toward High Point; it never arrived. One night in December 1913, a "mammoth airship" appeared in Wilson, shooting across the sky "like a meteor," hover-

Beachey prepares to give a flying lesson to E. Marie Sinclair, probably the second woman to fly in North Carolina. (Mrs. R. B. Middleton of New York had previously flown with Beachey at Pinehurst.) Courtesy of Mrs. Ellenore Eddy Smith.

ing overhead, and sending "blinding shafts of light from a powerful searchlight." A witness saw two men inside it, though another saw only a "paper balloon" with a candle that fell in a sedge patch and set his woods on fire. Hickory's February 1911 "cigar-shaped dirigible balloon" was possibly a breakaway from some distant mooring. Allegedly seen just five hundred feet above the corner of Hutton and Bourbonnais Streets, it was reported by only three people.[36]

Numerous teams operated on North Carolina's county fair circuit in the fall of 1911. Charles Witmer and Charles F. Walsh flew a Curtiss at the state fair in Raleigh in October and at New Bern and Winston-Salem in November, engaging, in Winston-Salem, in a mock battle with a Captain Wooten's Forsythe Rifles. Witmer flew at Charlotte in a driving rain; Harry LeVan, after entertaining Asheboro, almost crashed at Fayetteville and had to be replaced by K. Belton. J. B. McCauley's Curtiss delighted Elizabeth City in 1912, "the first time

# The Great Auto-Aeroplane Race
### PIEDMONT PARK 3:30 P. M. TODAY

A new Feature added. This is the same Aeroplane that beat the Pennsylvania's fastest special train from New York to Philadelphia; accepts the challenge of Cadillac, driven by Gilmer C. Thomas. A Race to the finish.

## Admission: Adults 50c, Children 25c, Grandstand 25c

Beachey's advertisement for an auto-airplane race at Winston-Salem in 1911. From *Winston-Salem Journal*, November 30, 1911, North Carolina Collection, University of North Carolina Library at Chapel Hill.

Exhibition pilot Eugene Heth about to take New Bern photographer Bayard Wootten on her first flight, November 1911. North Carolina Collection, University of North Carolina Library at Chapel Hill.

an airship," declared the local newspaper, "has ever been exhibited in Eastern North Carolina."[37] (The secretive flights of the Wright brothers and Harold M. Chase did not qualify as exhibitions.)

Disappointments plagued exhibitors. Flights scheduled at Waynesville in early October by airmen Gratz and Matlock were stymied by "wind from the west" and hills and trees to the east, evidently unanticipated phenomena. Wilkesboro's September show failed because plane and flier simply did not show up. At Wilmington in December 1912, Walter Brookins carried "the first air mail" (a feat claimed already by various airmen) in his Wright-Breguet, taking off with a single letter at Highland Park and dropping it at nearby Winter Park Gardens for "a designated recipient." A plan for Lieutenant Paul Cantwell of the Wilmington Light Infantry to shoot from the plane at ground targets was called off due to poor weather.[38]

The North Carolina shows had their share of crashes and mishaps but, fortunately, neither deaths nor serious injuries. A relatively mild climate may have helped, as well as the conservatism of fliers who knew that replacement parts and proper repair facilities were far away. The only recorded aerial death in the state in this era was that of Harold C. Brown, the Michigan balloonist who crashed at Lakewood Park near Charlotte in September 1911.[39]

A special treat were the flights of Thornwell H. Andrews at Greensboro during October 10–14, 1911. Andrews, a Charlotte native and skilled auto mechanic, was hired as a pilot by Lindsey Hopkins, founder of the Lindsey Hopkins Aviation Company. The interests of Hopkins, a rising entrepreneur, included Greensboro's Overland auto dealership and stakes in insurance, oil, cotton mills, banking, and other businesses, all of which made money. His airplane company sought to capitalize, while the fever still burned, on the country's new mania for aerial exhibitions.[40]

North Carolina's first professional pilot, the twenty-four-year-old Andrews, went to White Plains, New York, in the summer of 1911 to learn flying, while Hopkins visited Long Island to buy a plane. They were back in Greensboro in the fall, with "Curtiss biplane #12," in time to arrange their first air show, which opened on October 10 at the Central Carolina Fair. To heighten interest, Andrews, one of only about two dozen professional pilots in the country, was authorized to offer a $100 silver cup to any Tar Heel who would fly with him; it does not appear that any did so.[41]

The October 10 fairground flights were highly successful, despite a heavy rain, and continued at two flights daily until the fair closed three days later. "Every performance advertised," said the *Greensboro Daily News*, "was carried out to the letter"—a notable achievement in itself. Andrews was already regarded as "one of the most daring and successful aviators in the country." On the fourteenth, he made a spectacular flight over the city itself, at heights ranging from four hundred to three thousand feet, dropping advertising cards for local merchants. Aloft again at 4:00 P.M., he flew from the fairgrounds up Elm Street beyond the courthouse and then north over the countryside, his ninth flight in five days. He returned to his starting point after three or four miles and thirty minutes in the air. Crowds lined the streets and "every available window and the roofs of buildings" for the performance.[42]

His brain whirling with ideas, Hopkins made it known that he would buy a second Curtiss, learn to fly, and take it to San Francisco's Panama Exhibition. He sent Andrews on the road, cautioning him to take no unnecessary risks. The airman gave his next exhibition in the town of Lenoir, followed over a nine-month period by appearances in Georgia, Tennessee, Kentucky, South Dakota, Iowa, Kansas, Missouri, West Virginia, and Pennsylvania.[43]

Andrews had only two accidents in his entire career, the most serious at Gordon, Nebraska, in September 1912, when he had "a miraculous escape from death" and his plane was demolished. Hopkins, buying a second Curtiss, declared that "Thorny" was the kind of pilot who would fly even if he "had to use an advertising umbrella and a couple of palm leaf fans." Andrews never canceled a show "on account of mechanical trouble, wind, storm or weather." In his na-

Thornwell H. Andrews of Charlotte, North Carolina's first professional pilot, in the cockpit of a 1911 Curtiss biplane. From *Charlotte Observer*, June 12, 1932, North Carolina Collection, University of North Carolina Library at Chapel Hill.

tive Charlotte on May 12, 1912, he became the first person to fly an airplane over the city and was awarded a gold watch by its proud citizens. His Kansas City show on July 4—four flights "despite an unfavorable wind"—constituted, the mayor said, "the best exhibition of aviation ever seen" there.[44]

Hopkins and Andrews were the first Tar Heels to turn a profit with a flying machine, Andrews the first to make a living at it. His career, along with the brief enterprise of James N. Umstead, represented a notable contribution by his native state to the beginnings of exhibition aviation. Andrews bought the Curtiss from Hopkins and continued to hold exhibitions for some years. He died in Charlotte in 1934.[45]

Due in part to uncertainties of weather, machinery, accidents, and funding, Tar Heels saw fewer air shows in 1912 and 1913. Paul Studensky, a young Russian, flew at Gastonia in July 1912, and John G. Kaminski, at age sixteen a two-year veteran, at Lumberton and High

## A Suggestion.

How to Enjoy the Flight Tuesday Without Getting a Stiff Neck.

Cartoonist Billy Borne of the *Asheville Citizen* often depicted airplanes in his cartoons, like this one in 1911, a guide for watching air shows. From *Asheville Citizen*, August 12, 1911.

Point in 1913. J. S. Berger, H. Roy Wait, and Eugene Sodet (Sodet crashing through two fences and some buggies at North Wilkesboro) gave shows, as did DeLloyd Thompson at the state fair in 1914. But the novelty was wearing thin, and fliers had pretty well exhausted the tricks they could do with primitive machines.[46] Aviators would need better equipment and skills before they could prosper as before, which meant a sharp decline in shows until after World War I.

The European war was raging when a Wright seaplane performed at Wilmington in early 1915. Aviator Howard Rinehart promised that he would take off in his sixty-horsepower "hydro-aeroplane" on January 22–24 from the Cape Fear River at Riverdale Drive. Straining for novelty, he would bomb mock targets and race against a motorboat. He also came up with an impromptu stunt on the twenty-fourth, when, after flying downriver a mile and returning to the foot of Swann Street, his machine tilted sideways at about a hundred feet and dropped him into the river. The plane, badly damaged, was pulled to the wharf by a motorboat; Rinehart escaped with a good soaking.[47]

Despite the slackening excitement over airplanes on the eve of World War I, there was still a way to earn a living with them, and that was to jump out of them. A teenaged girl from Granville County, North Carolina, showed the way for the nation.

## THE DOLL GIRL

Georgia Ann Thompson was born on a farm near Oxford on April 18, 1893. She was so small at birth that her family nicknamed her "Tiny." As an adult, she stood four-feet-eight and weighed eighty pounds, so she kept the nickname all her life. Like most youngsters in her vicinity, Tiny worked at picking cotton and later took a twelve-hour-a-day job with a cotton mill. Married at fourteen, she was both an illiterate mother and a widow or grass widow in 1908, when she attended the James J. Jones Carnival at the fairgrounds in Raleigh.[48]

Of the attractions, nothing seemed half so exciting as the big canvas balloon, a gaily painted, ninety-two-foot colossus. Tiny watched transfixed as Charles Broadwick, its owner, did his free stunt of riding up a few hundred feet and parachuting to the ground.[49]

"Something told me that was what I wanted to do," Tiny said long afterward. "I just stayed there until he was done and told him I wanted to join up with him." Broadwick, learning of her infant and the recent death of Tiny's father, was dubious. He had lost his wife to ballooning three years before. Moreover, he already had Lola and Mademoiselle Teresa. But Tiny's petite form and fresh good looks would be something new, and Broadwick gave in. Securing her mother's permission, he added Tiny to his team.[50]

Women had flown in balloons since the 1780s; by the late nineteenth century many were professional parachutists. Over Asheville, on May 17, 1911, parachutist Florence Lorenza attempted a triple parachute jump, in which she was to cut the first two parachutes successively as she fell. This time, she cut the wrong one first and plummeted a thousand feet, with the weight of the other two parachutes speeding her fall. A sudden wind shift caused her to barely miss the post office roof and fall into some Patton Avenue electric wires, which saved her life.[51]

Years later Tiny Broadwick said that she made her first jump in Raleigh before the carnival left, though she had earlier stated formally that she made it on December 28, 1908, at Jacksonville, Florida. "All I can tell you," she said, "is that first jump was beautiful. I could see barns and trees and roads and people. Just beautiful. I felt such comfort coming down. I was in God's care. I knew I was going to be safe."[52]

Featured as "Miss Tiny Broadwick, the World's Most Daring Aviatrice-Parachutist," she accompanied the carnival across the country, changing her surname when Broadwick adopted her. In California, she performed night—as well as day—jumps, carrying flares and torches, especially at beaches on Sundays and holidays. Broadwick made the balloons and parachutes himself. Tiny, who came to be billed as "the Doll Girl," closed her mind to the dangers of her profession, focusing instead on its pleasures. This led to some nasty moments, as when she hit a mill coming down and broke an arm, dislocated a shoulder, and "banged up" an ankle. But the show went on at two jumps a day and $250 a week for the Broadwick team.

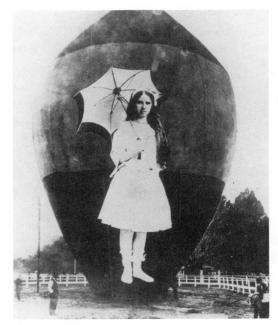

**Tiny Broadwick
depicted on a circus
balloon. She performed
as a parachutist as
early as 1908. North
Carolina Division of
Archives and History,
Raleigh.**

Very shy and too undereducated to enjoy banter with fans or reporters, Tiny was more at home in the sky, where she was a virtuoso
performer, than on the ground."[53]

Living in the Los Angeles area, Tiny found herself among a substantial community of military and civilian fliers. One day in 1912
while she was ballooning at Dominguez Field, a wind current carried her so far that an airman came in his plane to pick her up; it was
her first airplane ride.[54]

At the time there was still no parachute suitable for airplanes.
When planes crashed, the airmen often died. But army pilot and pioneer Glenn L. Martin was experimenting at Los Angeles with a trap
seat in which a parachutist could sit before dropping, and he asked
Tiny to test it for him. On June 21, 1913, with Martin at the controls,
Tiny, sliding from a small seat under the right wing and behind a
propeller, parachuted from two thousand feet and landed safely. She
was the first woman to jump from an airplane. (The first man had
jumped only fourteen months earlier.) Later she became the first
person to parachute from a seaplane into water.[55]

Tiny Broadwick prepares to make the first parachute jump from an airplane by a woman on June 21, 1913, in Los Angeles, with Glenn L. Martin as pilot. North Carolina Division of Archives and History, Raleigh.

It was easier to parachute from a plane than a balloon because the added height gave the jumper more control over where she landed. In 1914 Charles Broadwick designed a new type of parachute to strap on an aviator's back, and Tiny demonstrated it for officials of the army's Aviation Bureau. The idea, as with balloons, was for the chute to be opened by a cord fixed to the aircraft. On her fourth jump, the cord became entangled in the underside of the plane. Tiny, dangling 1,400 feet above Griffith Park airfield, cut the cord and snatched the remaining piece, subsequently known as the rip cord, to open her parachute. It jerked from the pack, making her the first person to perform a free-fall jump, more than five years before the first man would do so.[56]

Broadwick's parachute, with Tiny's impromptu alteration, gave aviators their opportunity to jump safely from a distressed airplane, a signal advance for aviation. But the army, fearing that its men would jump unnecessarily from balky aircraft, waited years before adopting the device. As a result, Allied pilots fought World War I without parachutes, and many young men died needlessly.[57]

Tiny jumped from balloons for eight more years, occasionally falling into trees, rivers, once even into a high-voltage wire, but never with serious injury. Regarded as the leading parachutist in the United States, she retired in 1922, after the novelty wore off and her act was no longer seen as spectacular. She was elected to the "Early Birds"—pilots who flew planes before December 19, 1916—though she was never a pilot. Tiny worked in an airplane factory during World War II and died in 1978, a grandmother honored in the aviation world. She was buried in Granville County.[58]

## HERALDS AND CASSANDRAS

An aeronautical culture was beginning to bloom across the country. Among North Carolina newspapers, the *Charlotte Observer* continued to speak to the most air-minded readers. The *Observer* was especially interested in introducing aviation to female readers and printed numerous feature articles for that purpose. In the fall of 1910 a syndicated feature included photographs of half a dozen women who had flown as pilots or passengers. All were ladies of high breeding and opulent lifestyle—Italy's Duchess of Aosta, France's Baroness de la Roche, American Mesdames Cortland Field Bishop and William K. Vanderbilt, among others. In April 1912 American drama critic Harriet Quimby made her flight across the English Channel.[59]

Anticipating Charlotte's first successful airplane exhibition in November 1911, the *Observer* featured women's aviation fashions. Mecklenburg County women might consider "the airplane hair bow," four yards of wide ribbon in a "great long looped bow . . . wired at the edges to keep in good form"—not unlike "the Flying Nun's" wimple. Soon came flying outfits, such as "a shirt and very full trousers" that could be converted into "a walking or automobile shirt," with a high-necked coat, the blouse "held snugly at the waist with a patent leather belt."[60]

Miss Grace McKenzie of Toronto, Canada, made news by conducting "the first aviation international romance" with airman Jacques de Lesseps, a reminder that aviation was "where the boys are." In late 1910 a biplane built by Miss Lillian Todd flew at Belmont Park, New York. A new genre of serial fiction, the aviation romance, began to appear in newspapers and magazines. Commercial advertising embraced planes and dirigibles as eye-catching icons for promoting everything from soap to Scotch.[61]

Editorialists gave considerable attention to two issues that troubled North Carolinians more each year. Between 1910 and 1914, airplanes made formidable progress in speed and agility, and pilots became more adept at handling them. Tar Heel journalists, reflecting the mood of the general public, found reasons for both dread and optimism in these advances. All alike wondered whether they were witnessing a march of glory or of folly. In June 1911 the *Charlotte Observer* reiterated Rudyard Kipling's admonition that nations of the

This Billy Borne cartoon, entitled "High Flyer," raps the high cost of bacon. From *Asheville Citizen,* November 7, 1910.

world must agree on protocols to prevent planes from evolving into the tools of ghastly new forms of war.[62]

The *Greensboro Daily News* was alarmed over airplane safety and reckless pilots. In thirteen months, 131 people had died in air crashes, mostly in France and the United States. The paper railed against pilots trying to outdo each other in sensational acrobatics or record setting. What was needed was "a period of repose, and a cessation of lofty and dangerous flights, until the problem of stability of flying machines can be in greater measure solved."[63]

The *Wilmington Morning Star* was appalled by the fatal crashes, on the last day of 1910, of Arch Hoxsey and Victor Moisant, two of aviation's brightest stars. But, it conceded, such daring revealed treacherous "air currents, or vacuums against which invention of the future . . . must take warning." Flights at extreme altitude, recently as high as 11,474 feet, though dangerous, were potentially valuable. "The sacrifice of human life is discouraging, but it may mean that science will get a new lesson that will yet turn to practical account."[64]

The *Salisbury Evening Post*, on the other hand, suggested that "the day is far distant, if it ever comes, when man will be using the air in safety like a bird." Daredevil flying did not prove "the value of the air ship to the commercial world." The editor advised that pilots not exceed a height of thirty feet. "Do this," he advocated, "until the behavior of the air ship comes to be more perfect. The danger to travel will thus be reduced to a minimum and there will be fewer fatalities and accidents."[65]

Aviation, though, showed signs of mindlessly overreaching itself. England's Claude Grahame-White, the *Observer* reported in late 1912, meant to fly the Atlantic in a hydro-aeroplane; the New England Aero Club was lectured on how the North Pole might be reached by airplane. But the prizes offered to those who performed such feats were a main source of Frenchman Louis Paulhan's thirteen-month income of $82,052, a princely fortune for the era. Such rewards ensured that no "period of repose" would be invoked, no thirty-foot ceiling recognized.[66]

Each month increased the prospect of a massive European war in which planes might play an important role. America must be ready to defend itself from the new menace. The *Wilmington Morning*

*Star* echoed Victor Loughheed's suggestion to Congress that two thousand planes could be built for the cost of a single battleship. But the *Star* proposed that world powers "cut the aeroplane out of warfare." Otherwise, it could only hope for a missile that would explode high in the air "and put the entire flock of flying machines out of business."[67]

In 1912 the *Greensboro Daily News* called attention to a recent aerial bombing test. A military plane, flying at one thousand feet, dropped two sandbags on a fort at Newport News, Virginia, both hitting near the ammunition dump. A race would doubtless ensue to learn whether artillery could be made to force planes up beyond their capacity to drop bombs accurately.[68]

The *Raleigh News and Observer*, however, hailed the progress of military aviation as "simply astonishing." In Los Angeles trials in early 1910, dirigibles had attacked a naval vessel at night. "What the future will develop," mused editor Josephus Daniels, "cannot be imagined by ordinary men. . . . The loss of a brave man . . . is a great price to pay; but in all discoveries . . . there must be sacrifices, even of life itself. . . . Many believe that, for light traffic, the airship will take the place of express trains." The *Salisbury Carolina Watchman*, with dead-on prescience, expected that aviation's pioneers would, "in a few years," perfect "a ship that will sail through the air with a degree of agility that will make it useful for transportation."[69]

For the present, there were other concerns. Northern property owners were outraged by spooked cattle and assaults on backyard privacy and neighborhood quiet. Connecticut home owners asserted property rights extending infinitely upward, an aim that might outlaw aviation altogether. Some communities banned Sunday flying lest the noise disrupt church services and days of rest.[70]

If war came, perhaps airplanes would be limited, as many supposed, to scouting missions; they would simply be a more mobile form of the balloon or dirigible. But suppose the airplane's destructive capabilities emerged in time for a European air war? What hellish nightmares might emanate from three-dimensional wars in which even the skies rained blood and ruin? In both Europe and America, legions of fresh-faced recruits were preparing to find out.

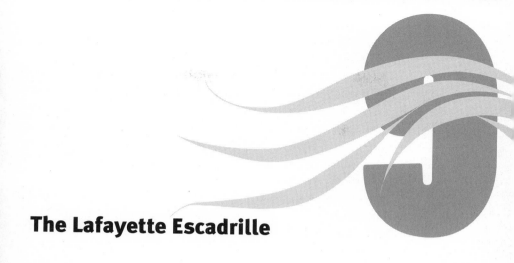

# The Lafayette Escadrille

Greensboro's Forrest E. Wysong was struck by the aviation virus in 1910, when a friend lent him a copy of *Aero* magazine. He bought a subscription, and, from diagrams in a book, the sixteen-year-old and his brother Paul, age fourteen, began building a twenty-foot-long biplane glider.[1]

So Forrest was already a precocious enthusiast when, in mid-1911, Lincoln Beachey performed at Greensboro's fairgrounds. The Wysong boys visited him at his hotel beforehand, and Beachey gave them passes and invited them to help set up his plane. After the exhibition, they met with him again and revealed their ambition not only to finish their glider but also to build an airplane. Beachey was impressed enough to spend several hours with them, sketching and discussing various types of planes.[2]

The Wysongs first made news that summer when the boys decided to pedal their bikes from Greensboro to Charlotte. Leaving Greensboro on June 21, they covered the ninety-nine miles to Charlotte in three days, staying overnight in Salisbury and then Lincolnton. The *Greensboro Telegram* noted that they were "not only cyclists, but future aviators. They are building a flying machine [the glider], at their father's manufacturing plant," where Forrest worked summers as a machinist. Olmedo C. Wysong, a manufacturer of woodworking machinery, was interested in aviation and indulged his sons' experiments to improve their mechanical skills.[3]

The glider, developed from diagrams in *Popular Mechanics*, was ready for trial in the fall, when Forrest was a freshman engineering student at the North Carolina College of Agriculture and Mechanic Arts (now North Carolina State University) in Raleigh. On the brisk morning of November 11, 1911, he took it to a hill at the outskirts of town, got airborne, and quickly crashed. "I didn't know what I was doing," he later admitted, "and I just stalled the thing."[4]

The father now apparently put a stop to his sons' airplane building. But Beachey, in response to a 1912 letter from Forrest, sent him plans for a pusher-type Curtiss biplane. Wysong secretly collected parts and found a shop on the college campus where he could assemble them. As Beachey directed, the elevator was "in the back, the way he had modified it."[5] Even with parts difficult and expensive to obtain as a result of world war, there remained young Americans who would stare down any obstacle to fly in a plane of their own.

Fellow student Gurdon L. Tarbox, a South Carolinian who had soloed in 1911, took an interest in the work, as did a pair of Wysong's friends from Greensboro. Tarbox not only could fly, but because his

**Forrest E. Wysong of Greensboro as a senior at present-day North Carolina State University in 1915. North Carolina State University Archives, Raleigh.**

father was an official at Curtiss, he could also obtain materials that were unavailable in Raleigh.[6]

Graduate student Stephen G. Bruner put up $450 to buy a second-hand 75-horsepower Roberts motor, gasoline and radiator tanks, and propellers. When finished in the spring of 1915, with Forrest soon to graduate, they took the plane to Raleigh's western outskirts, a site now occupied by Meredith College, where a farmer let them keep it "in a big hay barn."[7]

On March 16 Wysong made his maiden flight. "I ran it down the field . . . ," he wrote, "and jumped it off the ground a few times and finally decided, 'This is it, I'm going to fly it.' So I took off and went out over the trees." Too frightened to bank for a turn, he "made a great long sweeping turn over the field, just a few hundred feet high, and . . . looked back for the field. . . . It looked so small, just like a postage stamp. But I kept easing it down" in "roller-coaster descents until I got down." He soon made three more flights. They were the first in Raleigh by a native of the state.[8]

According to the *Greensboro Record*, the plane weighed 350 pounds, was 18 feet long, and had a 22-foot wingspan and propellers that were 8 feet in diameter. Its top speed was ninety miles an hour, and

**This cartoon of Wysong in his airplane appeared in his 1915 college yearbook. North Carolina State University Archives, Raleigh.**

it had recently stayed in the air for ten minutes at an altitude of 200 feet. Wysong boasted that he would fly it "from Raleigh to Greensboro in the best time" ever made between those two cities. Thereafter, the airplane would be kept in Greensboro. Beachey, the article noted, had died in a plane crash a few weeks before.[9] (Since he was hiding the project from his father, it is surprising that Wysong allowed the Greensboro paper to reveal his work.)

Immediately after the story appeared, "I got a summons," Wysong wrote, "to come to the [college] president's office, and when I walked in my father was sitting there." Olmedo and the president gave him a dressing-down and wrung from him the promise that he would not fly the plane again.[10]

But had he flown at all? On May 25 he was arrested on the complaint of his "angel," Stephen Bruner, for false pretense. Bruner said that he put up money because Wysong claimed to have worked with the Wright and Curtiss companies. The agreement was that Wysong would fly at fairs and elsewhere, dividing his profits equally with Bruner and students Harry C. Wharton and J. L. Hutton, who had each invested $500 in the project. But other newspapers contended that Olmedo intervened *before* the plane was tested.[11]

The court awarded the plane to Bruner. When he went to the barn to pick it up, he learned that it had been shipped to Greensboro, but he was able to have it returned. "Whether the machine will fly has not been determined," said the *Greensboro Daily News*, as only "running tests" had been made. Other papers referred sardonically to Wysong's "non-flying aeroplane."[12]

It is plausible that Wysong denied flying the plane to fend off his father, but at the time of the alleged flights the Raleigh newspapers made no mention of them. It seems doubtful that a plane, taking off and landing three miles from downtown Raleigh, could have escaped public notice in 1915. Wysong, however, became known in aviation circles as a man of probity and was later allowed, on the basis of his claim, to join the Early Birds. He even served a term as the club's national president.[13]

In the summer of 1915 Wysong went to work for Curtiss in Buffalo, New York, as designer, draftsman, and assistant project engineer. He later pursued a long, successful career in aeronautical engineering

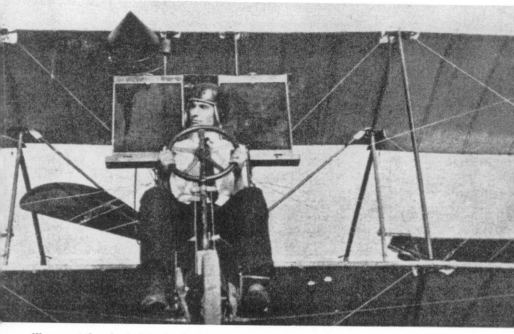

Wysong at the wheel of the Curtiss biplane that he built on his college campus in 1915. From *Greensboro Record*, undated tear sheet in Forrest Wysong Biographical File, National Air and Space Museum, Smithsonian Institution, Washington, D.C.

with the Douglas Aircraft Corporation. In World War I he served creditably as a navy pilot.

## INVENTING AERIAL WARFARE

The last great war, a contest between Prussia and France in the 1870s, had featured airpower in the form of hot-air balloons. Trapped for months inside encircled Paris, the French government ordered an emergency balloon-building program to stay in touch with its forces beyond and rally its citizens to greater efforts. Balloons that made it out of the city played a valuable role in keeping French armies in the field longer than might otherwise have been the case. The Japanese had used balloons to bomb Koreans, and the Turks had even used a plane to bomb the Greeks.[14]

At the onset of World War I, European powers saw the airplane as a tool for observation and reconnaissance; bombing was for dirigibles. Almost immediately, however, pilots were shooting at each other with small arms and, soon, machine guns. But engineers were working furiously on the challenges of this new form of warfare. It appeared that planes might help decide the outcome after all. In this competition, the United States, not yet a participant, lagged notoriously behind the belligerent powers. Neither politicians, nor generals, nor admirals on this side of the Atlantic grasped what was happening beyond. Even as it was being pressed to take sides, the U.S. government spent only grudging funds on aviation and had no plans to build pursuit planes or bombers.[15]

It was not for lack of ideas. Inventors pressed upon the government and military all sorts of contrivances for fighting the new warfare. In Wilmington, with its aeronautical heritage, the black headwaiter at the Orton Hotel, William D. Polite, saw the need, well before America entered the conflict, for a gun with which ground crews could knock planes from the sky.[16]

Polite, then in his early forties, was a native of Charleston, South Carolina, and partly of French ancestry, well educated, and "for many years" headwaiter at various Wilmington hotels. Ramrod straight in his "starched wing collar and cuffs, black tie, and frock tail coat," he was a devout Roman Catholic who headed a celebrated kitchen and dining room staff. During his career he trained over four thousand hotel personnel, mostly chefs and waiters. He was also an inventor, albeit not one who might be expected to contribute to military science. In 1911 he was granted a federal license to manufacture and sell "Polite's Pepper Sauce." The sauce, according to a Wilmington paper, was well received on the Lower Cape Fear market and was produced on a large scale, Polite later adding seafood and steak sauces to his line. His antiaircraft gun, nonetheless, was a sophisticated weapon and seemed well suited for wartime service.[17]

Polite moved to Charlotte and in 1915, still a headwaiter, hired a Washington attorney. After making inquiries at the U.S. Patent Office, the lawyer wrote Polite that three guns of the kind he designed

W. D. POLITE.

GUN.

APPLICATION FILED SEPT. 29, 1916.

1,218,458.

Patented Mar. 6, 1917.

3 SHEETS—SHEET 2.

African American William D. Polite of Wilmington and later Charlotte patented this antiaircraft gun in 1917. U.S. Patent Office, Washington, D.C.

had been patented but that Polite's carriage and gun were "in many respects" superior to the rest and patentable. Polite filed for his patent in September 1916 but had to wait until March 1917, just weeks before the United States declared war, to receive it.[18]

The Polite gun, capable of pivoting around on its carriage, automatically fed shells from its magazine onto a lifter that self-adjusted to receive each shell and place it in the gun's chamber. Its rapid-fire and flexible operation commended it for military use, but it does not appear that Polite found a manufacturer interested in producing it. He died in 1953, a revered figure and probably the first African American to design an antiaircraft gun.[19]

In 1915 Warrenton native William W. Christmas produced a design that dwarfed virtually every other heavier-than-air craft so far conceived. Late that year the *New York Times* carried a full-page feature on Christmas, who had recently moved his operation from Washington, D.C., to New York, outlining the nature and functions of his machine. Known as the Battle-Cruiser, the airplane, in development since 1912, was to be the dreadnought of the sky. Christmas pointed out that a limitation of the planes already in use was that they were built on the parallel-truss principle, that is, with wings bolted and wired to the frame rather than made part of the frame itself. The cantilever truss was an integral part of the frame, as in his earlier *Red Bird III*, allowing for far more capacity than any other plane.[20]

The European Allies, Christmas asserted, had already ordered eleven of his airplanes at $100,000 each. They would be 104 feet long, with wings spanning 180 feet, the largest planes ever built. Each of their "twin bodies" would be powered by a 1,600-horsepower motor driving a 20-foot propeller, giving it four times the power of any existing airplane. Each plane would carry from twenty-five to thirty-five bombs 5½ feet long and 14 inches in diameter, or up to 3,000 small "nitrous shells." A crew of six to eight, linked by telephone, would fly each plane and operate its two large-caliber, rapid-fire cannons and five machine guns, as well as its bomb racks.[21]

Each fuselage had its own pilot and dual controls; injury to one pilot or damage to one set of controls would not bring the plane down. If both pilots were killed, the crew could use automatic con-

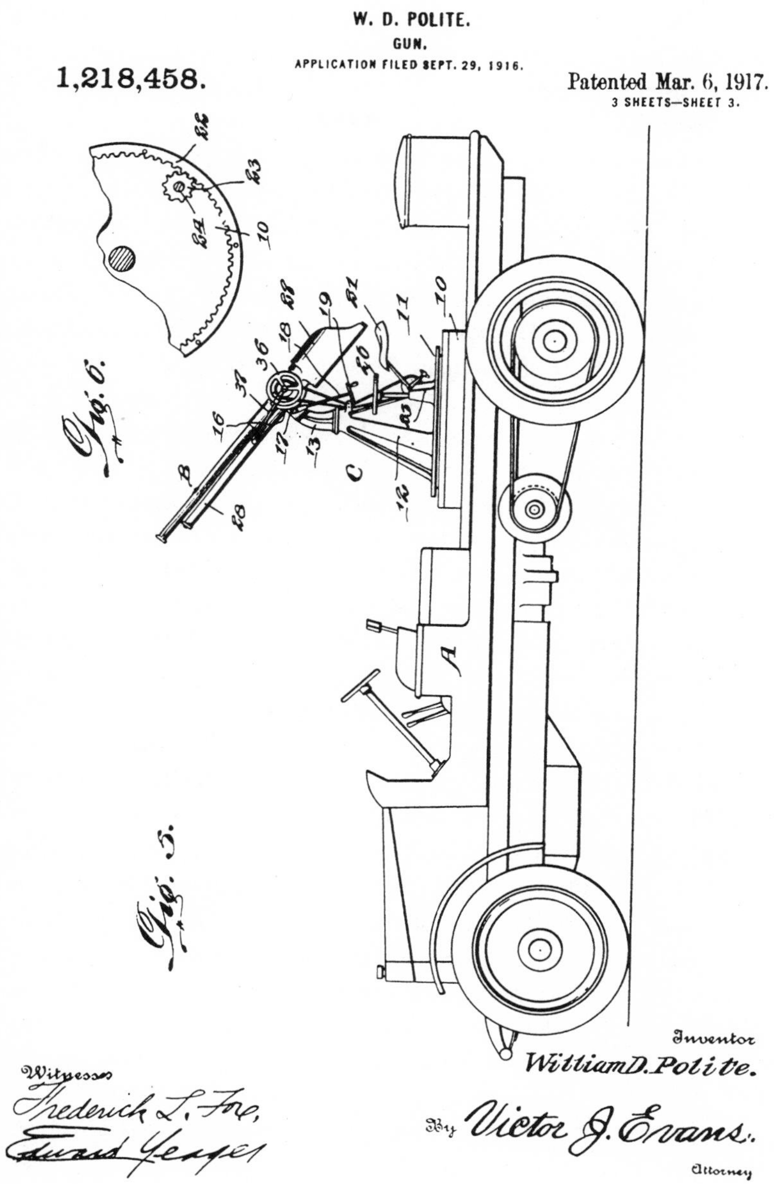

W. D. POLITE.
GUN.
APPLICATION FILED SEPT. 29, 1916.

1,218,458.

Patented Mar. 6, 1917.
3 SHEETS—SHEET 3.

Fig. 6.

Fig. 5.

Witnesses
Frederick L. Fox,
Edward Yeager

Inventor
William D. Polite.
By Victor J. Evans.
Attorney

This patent shows Polite's gun as a mobile weapon. U.S. Patent Office, Washington, D.C.

trols to steer it home safely. The aircraft would have a top speed of one hundred miles an hour, outgun any craft they could not outrun, and be used primarily to attack "moving bodies of troops." A hundred such planes, Christmas maintained, "could put any land or naval force . . . out of action in an incredibly short time." He was confident that his planes would bring the war to an early end.[22]

Given the builder's questionable reputation, it is not surprising that no more was heard of the Battle-Cruiser. No orders for it were placed. He simply sought to tease the army into funding his idea; it showed no interest. This did not, however, seriously undermine the Christmas Aeroplane Company—soon reborn as the Cantilever Aeroplane Company—or its controversial president.[23]

Advertisements in 1915 by the Christmas company, "Builders of the Largest Aeroplanes in the World," show a photograph of a modest-sized biplane with a peculiar fish-tale rudder, a tractor propeller, and skids as well as wheels. References in the ads to the company's "Armored Military Machines" indicate that these were warplanes, but this was only another mirage.[24] There is reason to suppose that by now Christmas had squandered his huge bequest from his grandmother and needed other resources to continue his work. Palmgren's *American* was another plane promoted without success for use by the military.

Other North Carolina inventors before and during the war included Olmedo C. Wysong. In 1917 Forrest's father designed a machine to aid in the manufacture of airplanes, and in the following year E. C. Gaskill of New Bern announced that he had invented an aerial bomb. Neither invention was awarded a patent or appears to have been produced. In 1916 the cruiser *North Carolina* was the first U.S. naval vessel equipped with an airplane catapult. Even when legitimate, however, advances in aviation technology were of little importance if political and military leaders could neither fathom nor anticipate their own needs.[25]

An important North Carolina influence on military aviation was the career of Josephus Daniels. In early 1913 Daniels, fifty-one-year-old editor of the *Raleigh News and Observer*, was appointed secretary of the navy by the new president, Woodrow Wilson. The office was a payoff for Daniels's work as publicity director of Wilson's campaign. As owner and editor of the *News and Observer* since 1894, Daniels had given aviation scant attention before the Raleigh air show of 1910 but warmed to the subject during the next three years. His views grew more favorable to aviation as his paper gained circulation and influence.[26]

As navy secretary, Daniels made an early visit to Annapolis. Arriving there on May 22, 1913, he inspected the buildings and grounds and was taken across the Severn River to the aviation camp. During his inspection of the five flying boats there, Lieutenant John H. Towers, the camp's senior pilot, invited him to go flying. Daniels stuffed some cotton in his ears and climbed into the Curtiss C-1, with Mrs. Daniels an uncomfortable witness. The plane pulled away from shore at 3:56 P.M. and flew a quarter of a mile downriver, toward the Chesapeake Bay. In about two miles, it rose to a height of five hundred feet, Daniels waving to onlookers below. Towers turned back for a leg of four miles upriver and then, returning, alighted smoothly on the river near the camp. The flight covered about eight miles in as many minutes. (In a later interview, Daniels expanded it to half an hour at eight hundred feet.)[27]

Daniels proclaimed his flight "delightful, I enjoyed the sensation thoroughly." He liked the smooth takeoff and found the view from above splendid and the pilot masterful. He felt no fear and, indeed, recalled times when it was "more dangerous to edit the *News and Observer* than to fly." He "returned to Washington more enthusiastic about the future of aviation."[28]

This was the first time that a cabinet member had flown in an airplane, and President Wilson, displeased, asked Daniels why he had risked his life; the secretary replied that he could not ask his men to do what he would not do himself. Postmaster General Garrard, said Daniels, "mildly called me a 'fool' for taking a flight." The secretary retorted that, before Garrard retired, the Post Office "would be carrying the mail in aeroplanes," which proved true. A week later Lieutenant Towers crashed the same plane into the Severn, killing a passenger and gravely injuring himself.[29]

Daniels had a formidable task ahead of him: to give the navy a credible aviation arm. After Towers crashed, that service had only "four unsatisfactory planes," four pilots, and no training station. But

AND NOT
A SIGN
OF THOSE
MISSING
PLANS

· Secretary Daniels Takes a Flier ·

**Cartoon of Navy Secretary Josephus Daniels in flight at Annapolis, Maryland, May 1913. He was the first member of a president's cabinet to fly in an airplane. From Josephus Daniels, *The Wilson Era* (1944), 1:291.**

navy planes had already proved their usefulness at Vera Cruz in 1914, during the Mexican Revolution, by warning U.S. warships of attack or ambush. In the following months Daniels secured from Congress "the first separate appropriation made for naval aviation"—$3.5 million. He took an interest in a young Virginia officer, Richard Byrd (later the first person to reach the North Pole by air), and gave him command of the secretarial yacht. The first American force sent to

France in 1917 comprised 7 navy fliers and 123 men directed to build naval air bases.[30]

Daniels may have been ahead of most of his admirals and fellow cabinet members in regard to the future of airpower. In early 1913 he believed that aviation would limit the duration and scope of war and aid in controlling the seas, "one of the elements contributing materially to the power and prosperity of a nation." Wars involving airpower would, in time, become "too destructive to be waged.[31]

In a speech at St. Louis, Missouri, in late 1913, Daniels predicted that the wartime role of navy aircraft would be not only scouting and surveillance but also "destructive operations against hostile fleets and fortifications." Aerial observers could follow the track of torpedoes, water seen from above being clearer than from the surface, and accurately direct naval artillery. Planes and dirigibles could detect mines better than ships could, yet a plane cost no more than a torpedo, an armada of planes less than a single battleship. He regretted that the United States was far outdistanced by the British, French, Italians, Germans, and Russians in military aviation. A German dirigible with wireless radio, four machine guns, and a one-pounder cannon could carry seven tons of explosives for up to 375 miles. The French air force, Europe's largest, was said to have six hundred planes and a huge appropriation for 1914.[32]

Granted that America was not yet endangered, Daniels noted, if a threat came, the nation had only twenty-four military planes (seven of them owned by the navy) and ten on order. Given the need to guard the new Panama Canal and other distant bases, the U.S. program was pathetically inadequate. One reason, he felt, was that Americans were losing interest in aviation since the rush of achievements from 1909 to 1912. "The days of golden harvest of exhibitions," he said, were over; "the manufacturing industry lies prostrate" for lack of support, and aeronautics "is limited to a few enthusiasts."[33]

Aviation's deficits, Daniels urged, must be addressed by multiple means. Technical schools should teach flying and foster aeronautical science. A proposed naval aviation reserve should be approved. Public pressure on Congress might be stoked by a movement to honor the deceased Wilbur Wright with a building at the Smithsonian Institution to house "the first machinery and instruments" of flight. Daniels meant to bring the level of the navy air service to that the U.S. fleet enjoyed among the world's navies.[34]

Yet Daniels' actions did not always correspond with his hopes. For him, as for his predecessors, the navy primarily meant ships. So, for all his good intentions, the fleet entered World War I with puny air support. Three days after his St. Louis speech came a tragic reminder of what was wrong with the country's aviation program: the death of a North Carolina airman in a military crash.


## WARBIRDS BEFORE THE WAR

In 1914, after more than a decade of development, airplanes still seemed less likely than dirigibles to provide important public services. Nor were civilian companies able to suggest constructive applications of airplanes. The delivery of mail, let alone passengers and produce, was not yet widely seen as a practical use for aircraft.

It was a mixed blessing that heavier-than-air flight would soon alter military training and maneuvers in ways that would transform warfare. It was an advantage because nations at war would pay any price to make their planes better than those of their enemies, thus compressing into three or four years the progress that might otherwise have taken decades. It was a curse because, having found acceptance as a weapon, airplane development was subsumed under the requirement of speed, which, as exemplified in small open-cockpit biplanes, had little obvious application to foreseeable peacetime purposes.

The army, which bought its first plane from the Wright brothers in August 1909, was in scarcely better shape than the navy. Wilbur Wright was the army's first instructor, and the new Signal Corps air service (the ancestor of the air force) had commissioned its first pilot in May 1912.[35] But the army planes did not appear to warrant much enthusiasm for airpower.

Pancho Villa furnished the army with a test of its air arm in 1912, when his rebels began making raids into New Mexico. In February 1913 the army sent seven pilots and nine planes to Texas to serve with ground and naval units, and on March 5 it organized the First Aero Squadron at Texas City, near Galveston. Among the squadron's seven members was Lieutenant Eric L. Ellington of Clayton, North Carolina, who commanded one of the unit's two companies.[36]

Ellington followed a circuitous path into the cockpit of an army plane. A son of Johnston County sheriff and former Confederate officer Jesse T. Ellington, he was born in Clayton in 1889 and at age sixteen won appointment to the Naval Academy. After completing his course work in 1909 as third in a class of 309, he was assigned to an Atlantic battleship to learn practical seamanship. In about a year he was commissioned lieutenant and served as navigator on the destroyer *Lawrence* for a 1910–11 cruise around Cape Horn to the Pacific.[37]

Owing to chronic seasickness, Ellington in 1911 applied for transfer to the cavalry, a move that only the chief executive could authorize. President William Howard Taft consented, and in September Ellington transferred in grade as a second lieutenant to the Third Cavalry at Fort Sam Houston, Texas. Within a year, at his request, he was assigned to Signal Corps aviation. After training at Marblehead,

Lieutenant Eric L. Ellington of Clayton, 1889–1913, a member of America's first military squadron. Courtesy of Ellison Library, Smithfield, North Carolina.

Massachusetts, College Park, Maryland, and Palm Beach, Florida, he won his wings in February 1913 and was with the first group sent to Texas City, 325 miles from the Mexican border.[38]

Fortunately, there was no fighting, and the inexperienced First Aero Squadron was able to concentrate on training with its machines. In June Ellington was sent to the North Island Signal Corps Aviation School, near San Diego, to train pilots. There he found a dozen or so pusher-type Wright C-1 and Curtiss biplanes, the bulk of the army's air arm. Three previous C-1s all had crashed, killing their pilots, and two more soon did so. The C-1 was said to be dangerous to fly, the Curtisses scarcely any better.[39]

Ellington, the ace of his squadron, was named chief instructor at North Island, but in October he was able to return home to Clayton for a brief visit. On his departure for California, his sister Lucille asked him if he would consider leaving aviation. He replied that flying was "the greatest thing of the present day and somebody has to give their life for it. If I knew I was going to die, I'd still keep doing it."[40]

On Friday, November 20, San Diegans witnessed a splendid exhibition by Ellington as he flew around the area for forty-three minutes at a top altitude of 3,500 feet. At 7:25 A.M. on Monday the twenty-third, three days after Secretary Daniels's call for a greater effort in naval aviation, Ellington and student pilot Hugh M. Kelly took side-by-side seats on a dual-control, fifty-horsepower C-1 for a training flight. With Kelly at the stick, they rose about 300 feet into a hazy sky with a light northwest breeze. Near the southern point of the island, they turned back toward the field. The six-cylinder engine was missing badly.[41]

A mile short of the hangars and two hundred feet up, the pilot throttled the motor down for a glide, evidently meaning to make an emergency landing short of the field. Suddenly, said witnesses, full power was again turned on as Ellington, seizing the controls, tried to overfly some bushes, his angle of glide steepening into a headlong dive. There was a sickening crash, the motor slamming forward into both men; Kelly's body was driven into the ground up to the waist. Both men, badly crushed, died instantly.[42]

An army board concluded that the cause was "stalling due to remaining too long in air with missing engine and starting glide"—in a word, pilot error. Lincoln Beachey, who was in San Diego for an exhibition, witnessed the crash and angrily disputed the finding. In a front-page statement in the November 25 *San Diego Chronicle*, he lashed out at the "parsimonious policy" that led to the "absolute slaughter of good, game men who volunteer." The policy forced airmen to train in "antiquated machines or fly nothing at all." He had seen mechanics at North Island repair planes with scraps found around hangars. The airplane, he insisted, was at fault in Ellington's crash. The outmoded, underpowered c-1 was "practically the same type used by Orville Wright four years ago." Its pusher propellers, when throttled, threw a blast of air backward, creating a whirl under the elevator, forcing it up, and causing the plane to pitch forward.[43]

Beachey promptly received telegrams from War Secretary Lindley M. Garrison and Navy Secretary Josephus Daniels inviting him to Washington to discuss problems with aviation training and equipment. In doing so, he helped persuade the army to make some long-delayed reforms.[44]

The seven men killed in army crashes during 1913 brought the total to thirteen since 1908. Embittered pilots, in what was described as "an insurrection" because they bypassed normal channels, sent a list of demands for training changes directly to Brigadier General George P. Scriven, head of the Signal Corps. Most were adopted and the army decided to order no more pusher-type planes.[45]

The changes fell far short of what was needed. By 1914, the fifth year of military aviation, there were still only fourteen qualified army pilots. Larger, more powerful Burgess and Martin planes were ordered, but the tiny service suffered three more air fatalities in the next two years. Chronic small budgets meant that neither army nor navy had any American combat planes to send to Europe during the whole of World War I.[46]

Ellington's body was shipped east and buried in the family plot in Clayton. His funeral was conducted on December 2, 1913, with full military honors, including a twenty-one-gun salute fired by ten men from Fort Caswell. Secretary Daniels sent a floral tribute. A quartet sang "Some Day We'll Understand." Soon afterward, an army airfield south of Houston was named in Ellington's honor.[47]

New raids by Pancho Villa into New Mexico persuaded the U.S. government in March 1916 to send forces under General John J. Pershing to run him down. To these were added eight planes of the First Aero Squadron, new Curtiss JN-2s or "Jennies." The Jenny, with its long, narrow fuselage and staggered wings, was superior to any of its predecessors. Captain Benjamin Foulois headed the eleven pilots sent to fly them on patrols, take dispatches to and from distant units, and similar missions.[48]

Other air units went to Brownsville, Texas, and points near the Mexican border. Among them was the Ninetieth Aero Eagle Squadron led by Lieutenant Hubert McLean of Gibsonville. McLean had received his first flying lesson in Greensboro from Lincoln Beachey in 1911. Enlisting for army aviation in 1915, he trained in New York and at Kelly Field, Texas. He was then ordered to Del Rio and Eagle Pass, Texas, fields on the Rio Grande River and the Mexican border.[49]

Serving under McLean at Eagle Pass was Lieutenant James "Jimmie" Doolittle, a young man destined for an illustrious aviation career. Seasbrook B. Renn, of Raleigh, another Tar Heel who served in the Southwest before America entered World War I, joined Aero

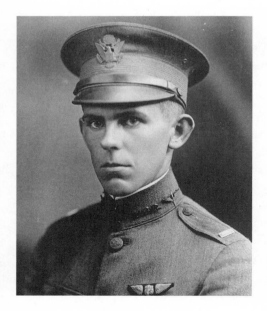

Lieutenant Hubert McLean of Gibsonville, who flew on the Mexican border in 1916 against Pancho Villa. Courtesy of Charles E. Moore.

Squadron Number One at Columbus, New Mexico, in 1916, when the army still had only five operational squadrons.[50]

Through no fault of the pilots, airplanes achieved little during the Mexican troubles. The Jenny was underpowered for the duties expected of it, including flights over tall mountains. Moreover, it took only about six weeks' service to wear out a plane, and many were lost to accidents or lack of spare parts. Those remaining were eventually recalled to fields a good distance from the fighting.[51]

In 1915 an act of Congress authorized each state to form four "naval reserve aerial crews" of six men and two officers each, the government to provide an airplane for each crew. In North Carolina there was talk of creating such a unit at Wilmington. Captain C. D. Brabham of the state's naval reserves invited Harold M. Chase, the reserve officer who had hopped Gouverneur's quadriplane along Shell Island beach in 1910, to pursue the matter. Chase conferred with other officers, but nothing seems to have come of it, probably for lack of an airplane. An army air reserve was formed, but its records were lost and little is known of it.[52]

At the start of 1916, the army had twenty-five planes and twenty-

Lieutenant McLean (fourth from right) stands with members of the squadron he commanded in Texas in 1916, including Jimmy Doolittle (second from right). Courtesy of Charles E. Moore.

three pilots, though World War I was already eighteen months old and likely to last years longer. The situation with Mexico was settled quickly after the United States was drawn into Europe's war in April 1917. Hubert McLean stayed in the army to become an instructor and, after the war, a commercial pilot before his death in a Mexican plane crash in 1929.[53]

Others destined to fly in combat enrolled before America entered the war in so-called riving (or opening-up) squadrons used to survey logging roads in the Pacific Northwest. There, in Washington State's "Big Spruce Forest," most of the wood used to build military airplanes was sawed. Pilots like Leon W. Powell Jr., of Warrenton, who enlisted in early 1917, helped locate timber and keep roads open to trucks hauling airplane spruce. Powell was later a pilot in the war overseas.[54]

War propaganda was already making North Carolinians jumpy. In September 1915 "strange aircraft . . . presumably of a very powerful type" were reported over Pollocksville, in Jones County. Alex McDaniel and William Coston were among those who saw them heading west in late evening, casting searchlights, and heard them pass eastward early the next morning. In February 1917, with America about to enter the war, residents of Stoney Creek township in Wayne County watched five planes "for more than an hour" circling over the Neuse River and "believed they were commanded by German spies."[55] Germany never revealed what it sought in Jones County or why.

## FLYING FOR FRANCE

Some Americans, especially idealistic lovers of French history and culture, went to war more than two and a half years before the United States became a belligerent. The first two were brothers Kiffin and Paul Rockwell, sons of a prominent Asheville family. They were working in Atlanta when France declared war on Germany, August 3, 1914. Before the sun set that day, Kiffin wrote to the French consul in New Orleans volunteering Paul and himself for military service.[56]

Kiffin Yates Rockwell was born in East Tennessee in 1892, his father a North Carolina native who died a year later. Kiffin's mother, Dr. Loula Ayres Rockwell, moved with him and Paul, then seventeen, to Asheville in 1906. (Novelist Thomas Wolfe, a younger Asheville contemporary, wrote in *Look Homeward, Angel* that Kiffin was considered "a sissy.") After attending high school in Asheville, Paul A. Rockwell enrolled at Washington and Lee College and Kiffin at the Virginia Military Institute (VMI) in 1908. Within a year, Kiffin received an appointment to Annapolis but declined it because he craved action and feared that the navy would not see it in his lifetime. Six feet four, with deep-set blue eyes, he had many friends, was a fine dancer, and charmed the ladies.[57]

But Kiffin had no clear goal in life. After visiting Paul at Washington and Lee several times, he decided in 1911 to leave VMI and enroll there. Both young men turned to journalism; at the time, Washington and Lee had the country's only journalism school. Before long, Kiffin left without graduating and set off for the West Coast with thoughts of settling there. He worked briefly in San Francisco but was back in Asheville in late 1913. Meanwhile, Paul graduated and became a reporter for the *Atlanta Constitution*. Kiffin joined him in Atlanta and took a position with an advertising agency.[58]

The Rockwell brothers grew up with a love for Gallic civilization. Without waiting for a reply to their letter to the French consul, they left Atlanta on August 6, 1914, for New York, not even stopping in Asheville, to book passage for Europe. The next day they were on the American liner *St. Paul*, sailing for Liverpool. Both felt that the issue to be decided in the war was that of representative government versus authoritarian rule. They had no doubt as to their moral duty: to defend democracy. On reaching Paris, they found themselves barred by their nationality from the French army, so they enlisted on August 30 in the multiethnic French Foreign Legion. Several other Americans living in France and England when the war began soon joined them; it was an eager lot. Grilled by French officials as to their prior military experience, one volunteer cited five years' service in the Salvation Army. The Rockwells claimed to have fought in the Mexican army, their awkward marching being the result of employment as guerrilla irregulars.[59]

Paul, who was injured at the front in late 1914, was unable to return to combat and worked as a correspondent. Kiffin continued to fight in the trenches until May 1915, when he received a severe thigh injury at a battle near Arras and had to crawl back across the battlefield to find help.[60]

Meanwhile, Legionnaire William Thaw, a Pennsylvanian who had learned to fly in America, began talking with the Rockwells and others about forming an air unit of their own. In early 1915 they sought the permission of the French government. At first the authorities were not interested, having as many French volunteers as they needed. Eventually, however, they concluded that such a group might help galvanize American public opinion and bring the United States into the war. On March 3, 1916, seven Americans, including Kiffin, were allowed to form an American squadron. This group, enlarged in the following weeks, went by several names but was best known as the "Lafayette Escadrille." Paul, an honorary member, acted as the unit's historian and publicity agent.[61]

After just four weeks of training, the group, now including James R. McConnell of Carthage, North Carolina, was sent to Luxeuil-les-

Bains, an Alsatian resort town near the German border, for practice in obsolescent planes. The arrival at Luxeuil on May 14 of the latest 100-mile-an-hour Nieuport-17 pursuit planes signaled that the Escadrille was considered ready for war.[62]

Jim McConnell, who was born in Chicago in 1887, was, at twenty-nine, among the older members of the Escadrille. As a teenager, he drove a car from Chicago to New York, a headline feat. At the University of Virginia, he played football and founded the school's first "Aero Club" but left in 1910 without graduating. Like Kiffin, he had no settled direction. He briefly sold cars in New York before joining his family, then living in Carthage, where his father was vice president and general manager of the Randolph and Cumberland Railway. Jim became the railroad's industrial and land agent and secretary of the Carthage Board of Trade, writing promotional material for the Sandhills region. He was "actively interested," wrote the *Raleigh News and Observer*, "in the development of his adopted section and . . . through his articles many were brought in closer touch with Moore, Hoke, and adjacent counties." He especially boosted grape and small fruit culture by Sandhills farmers.[63]

In January 1915 McConnell left for France and joined the volunteer American Ambulance Corps. As a driver, he won the Croix de Guerre for bravery under fire but, aching for combat, enrolled in the Foreign Legion in October. During this time he continued to write, now as a war correspondent for his hometown paper and, later, for such national magazines as *Outlook* and *The World's Work*. After falling in with Thaw's group, he was allowed to transfer in March 1916 to the Lafayette Escadrille, then just organizing. McConnell was a valuable addition, having lived in France for a time before the war and speaking fluent French.[64]

At Luxeuil, the Americans were housed in a lovely villa, well staffed with Oriental servants, adjoining the town's hot springs. Everything was new, from offices to Fiat trucks. There was a cadre of seventy-odd drivers, clerks, mechanics, motorcyclists, stretcher bearers, and others. Thousands of Indo-Chinese laborers cleared land, laid railway tracks, and so on. Two cars with brass headlamps stood by to carry the fliers between villa and airfield. They dined in style at the best hotels and were treated handsomely by civilians. To McConnell, it recalled "the ancient custom of giving a man selected for the sacrifice a royal time of it before the appointed day."[65]

On May 16, 1916, the Escadrille made its first patrol over enemy lines, flying in tight formation across the German border, sixty miles east of Luxeuil. On this and succeeding days, they saw few enemy planes, owing to the fact that the Germans were concentrating forces on the Verdun front, one hundred miles north, in the heaviest campaign so far. The few enemy planes that were spotted quickly sought the cover of their antiaircraft guns, so the first four patrols passed without incident.[66]

On the Escadrille's fifth patrol, May 18, Kiffin Rockwell was flying alone in his silver Nieuport at about a thousand feet when his motor began missing. As he turned toward Luxeuil, a German two-seater observation plane came out of a cloud below him. Ignoring his motor, Rockwell sped toward the German, who went into a steep dive and headed, as usual, back toward his antiaircraft batteries. But,

James Rogers
McConnell of Carthage
in France, 1916. North
Carolina Division of
Archives and History,
Raleigh.

*The
Lafayette
Escadrille*

211

on this occasion, the German abruptly reduced speed, apparently inviting engagement.[67]

Rockwell, six hundred feet above and half a mile behind, had a decided tactical advantage and nosed down to come in under the German's blind spot. The enemy gunner in the rear cockpit opened up with his machine gun, hitting the Nieuport with one bullet. Rockwell continued on through a hail of shots, a trait that would characterize all his later battles. At about thirty yards, he got off a burst of four shots before his gun jammed and he swerved sharply to avoid a collision. As he passed the German plane, he turned to come in again from above, hoping to get another burst from his gun; but he saw in a moment that it was needless. The enemy pilot was crumpled sideways in his seat, and his gunner had fallen backward, also apparently dead, his gun pointing straight up. The crippled plane, trailing smoke, went into a gyrating dive and smashed to earth just behind some German trenches. Coming in low over the site, Rockwell saw the wreck in flames before heading his own sputtering craft home.[68]

Kiffin Rockwell was the first American airman and the first member of the Escadrille to down an adversary, a feat lauded in a "Strange as it Seems" cartoon for doing so with the first four shots he fired in combat. Paris cheered him, and brother Paul sent a bottle of old Bourbon, which the Escadrille dubbed "the Bottle of Death," each man to drink from it only after downing an enemy. (It was soon empty.) Rockwell wrote home of his victory: he felt "very good over it" and was to receive the Medaille Militaire and Croix de Guerre.[69]

The brief era of high living and rare encounters with German aircraft ended within forty-eight hours of Rockwell's triumph, when the Escadrille moved to the raging Verdun front northwest of Luxeuil. Based eighteen miles south of Verdun, the Americans on May 21, 1918, found themselves abruptly in the thick of war. Several of their planes were shot up in the first day's action, but no pilot was seriously hurt. On May 25 and 26 the unit shot down three more Germans.[70]

The Lafayette Escadrille enjoyed the best press coverage of any squadron in the war. The pilots were mostly from well-to-do or, like the Rockwells, at least well-educated families. They tended to be

**Kiffin Rockwell in France, 1916. He was the first American to shoot down an enemy plane. North Carolina Division of Archives and History, Raleigh.**

highly literate, perceptive, and especially well placed to bring the war home vividly to those distant from the fighting. Several were gifted writers who, with publicist Paul Rockwell, sent batches of articles to periodicals in western Europe and America. No one abreast of the war could miss the accounts of exploits that poured from their pens. A flamboyantly devil-may-care band, few of them shied from publicity or scorned celebrity. The Escadrille—with its Sam Browne belts, Savile Row boots, shipboard tunics, and swagger sticks; its sense of high purpose; its enviable flying skills; its swashbuckling lust for life; and its lion cub mascot "Whiskey"—was frantically admired. Its very existence was a potent argument for American cooperation with the Allied forces.[71]

From Paris, Paul Rockwell wrote regularly for the *Chicago Daily News*, as a freelancer for English and French papers and magazines, and as a correspondent for the French army's Grand Headquarters. He later edited a volume of Kiffin's letters and was the author of *American Fighters in the Foreign Legion, 1914–1918* (1925). Jim McConnell, during a two-month convalescence from injuries, wrote *Fighting for France* (1917), which was an instant classic. Other books

*The Lafayette Escadrille*

213

**Les Boulevardières. Paul (standing) and Kiffin Rockwell in Paris, 1916. North Carolina Division of Archives and History, Raleigh.**

by members of the group included *I Flew with the Lafayette Escadrille* (1953) by Edwin C. Parsons and *The Lafayette Flying Corps* (1920), coedited by James Norman Hall.[72]

In the Verdun sector, Kiffin Rockwell distinguished himself, in the words of a comrade, as "a demon in the sky." He flew up to four exhausting sorties a day, in July and August engaging in an astonishing seventy-four fights, more than any other pilot flying for France. In this period he reportedly shot down seven German planes, all but one so far behind enemy lines that no one could verify them.[73]

On May 24, while returning from eight dogfights with low fuel, he fell in with a comrade, Victor Chapman. Immediately observing a formation of seven German aircraft below, they attacked. Rockwell, according to one account, "used his usual tactics of waiting till he was almost on the point of collision before firing," as Chapman chased another plane. As Rockwell closed on his target, an illegal "explosive bullet struck one of the wing-supports near his head, spraying deadly slivers of wood and lead." A piece creased his skull, another his forehead. Wiping blood from his eyes, he "let fly at his enemy: one short burst at seventy meters and another at forty." The German plane trembled, stopped, and exploded. "The whirling body of the pilot" passed within twenty feet of Rockwell's plane.[74]

"Practically out of gas," adds the account, "weakened from the loss of blood and shock from his wounds, his ship a perfect sieve," Rockwell turned for home. Though all but blinded by blood, he made it to his home field, had his wounds dressed that evening and, with his head swathed in bandages, went on patrol the next morning.[75]

In July 1916 the Allies opened an offensive along the Somme River, far to the west, drawing many German units away from Verdun. By the time the twelve men of the Escadrille were moved briefly to Paris in September, they counted a dozen verified kills against the loss of only one man: Victor Chapman on June 23.[76]

The Escadrille soon returned to Luxeuil, where it found two British pursuit squadrons. The Germans had reinforced their own air units in the area, and dogfights were numerous on this formerly sleepy front. The Americans waited for new Nieuports, Rockwell and the squadron's ace, Raoul Lufbery, drawing the first two to arrive. On September 9, at three thousand meters above Vauquois,

Rockwell scored his fourth accredited victory, killing the plane's observer with his first shot but retreating quickly when he was attacked by two more German aircraft.[77]

On September 23, 1916, Lufbery and Rockwell set off on patrol but became separated. As he neared the lines, Rockwell spied an enemy observation plane below and, in his patently suicidal fashion, roared in for the kill through the fire of two machine guns. At the last moment, almost colliding with the German, he got off a burst of fire. But he took an exploding dumdum bullet in his chest and throat and died instantly. As his plane nosed downward, the wings on one side fell away. He crashed behind French lines at a spot two and a half miles from where he had downed his first German plane in May. Photographs of his ravaged face and body were sent to Geneva in protest over the German use of exploding bullets, but no action was taken. It was Rockwell's 142d aerial battle, more than all the rest of the Escadrille combined. He was the second American airman to die in the war.[78]

The Germans turned their artillery on the wreck, but Allied soldiers ran forward and retrieved Rockwell's body. His funeral at Luxeuil was a memorable one, with a regiment of French territorials attending, along with Escadrille comrades, eight hundred personnel of Britain's Royal Flying Corps, and hundreds of French pilots and mechanics. Circling planes showered flowers from above. Captain Georges Thenault, leader of the Escadrille, observed: "When Rockwell was in the air, no German passed . . . and he was in the air most of the time. The best and bravest of us is no more."[79]

Kiffin Rockwell received France's highest military award, the Medaille Militaire; the Croix de Guerre with four palms and a star, representing five citations; and the Chevalier's Cross of the Legion of Honor. He also won the Verdun Medal, Combatant's Medal, Volunteer Combatant's Medal, and Croix de Blesses (France's Purple Heart). As in the case of Eric Ellington, a military airfield, this one in San Diego, was named in Rockwell's honor.[80]

At the time of Rockwell's death, his fellow Tar Heel Jim McConnell was lying in a hospital bed nursing a back injury. As fate would have it, he had had poor luck from the beginning of his combat career. At Luxeuil in May 1916, he wrecked his plane just before the Escadrille left for Verdun and had to wait for a new one before rejoining his unit. On his first flight at Verdun in June, he brashly attacked six German planes returning home from a bombing mission. When his gun jammed at the first shot, the Germans riddled his plane mercilessly, putting nine holes within a foot of his head. Limping home, he had to wait more weeks for a replacement plane, flying only when someone was not using his own machine.[81]

In August McConnell and four others were patrolling over a ground fight near Fleury to keep German observation planes away. Since several enemy aircraft waited behind their lines for the Escadrille to leave, the Americans stayed until dark. On the way home, McConnell's motor began to miss, and he descended to look for a landing site in unfamiliar country. On the way down, his plane hit a telegraph wire and crashed. His back was badly sprained, but he refused hospitalization and was flying again as soon as he got a new plane. "Finally," says Paul Rockwell, "he was unable to move and had to be taken away, greatly to his sorrow and disgust."[82]

*Flying for France*, dashed off during McConnell's hospital stay, created a wide audience for his work. In a characteristic you-are-there passage, he describes his experience of war above the trenches: "During heavy bombardment and attacks," he wrote, "I have seen shells falling like rain. The countless towers of smoke remind one of Gustave Dore's pictures of the fiery tombs of the arch-heretics in Dante's 'Hell.' . . . Now and then monster projectiles hurtling through the air close by leave one's plane rocking violently in their wake. Airplanes have been cut in two by them."[83]

Back in action in October, but not yet healed, he was again removed from the front, returning only on March 10, 1917, though friends begged him not to. Three days later, Paul Rockwell received a letter from him describing how, flying above the lines a few days before, he had landed in a field with a dead motor. By now he was quite discouraged and felt "rheumatism not only in my back but in my knees." Like the rest of his squadron, he had become fatalistic about whether he might die sooner rather than later: "My burial," he wrote to his comrades, "is of no import. Make it as easy as possible for yourselves. I have no religion and do not care for any service. . . . God damn Germany and Vive la France."[84]

At 9:00 on a misty Monday morning, March 19, 1917, McConnell and Edmond Genet left in Spads to cover French observation missions over German lines near the village of Ham. They hovered overhead for about an hour before McConnell turned north across the German lines, his purpose obscure. Genet followed at an altitude of three thousand feet and soon saw two German planes at a higher altitude. McConnell apparently did not see them, but Genet throttled to full speed to reach seven thousand feet, the level of the nearest enemy plane. The Germans opened fire, riddling Genet's upper left wing. For a few moments, the planes jousted, but then his opponent, perhaps not aware of Genet's plight, flew off to the north. Genet dropped to three thousand feet looking for McConnell, but, seeing nothing of him, returned home hoping to find him there. He had not returned. Late in the day, Lufbery went to look for him but found nothing.[85]

Three days later, French cavalry reported having seen McConnell engage two Germans. His plane had fallen into an apple orchard behind enemy lines. The next day, French troops found a wrecked Nieuport near Petit Detroit with the pilot's body inside. McConnell's captain went to the site and found that enemy troops had taken his valuables, even his boots, and left him unburied. He was not credited with a kill during his brief flying career.[86]

McConnell's body received a hero's funeral in Paris. Gutzon Borglum, the sculptor of Mount Rushmore, was commissioned to execute a winged figure in bronze in his honor for the University of Virginia. Friends and admirers in Carthage erected a memorial shaft at the courthouse. His tomb was built, by members of the Sixty-fifth French Infantry, with stones from streets and buildings at Villequiers, where he died.[87]

McConnell was remembered in the Escadrille for his literary and artistic taste, his love of sports and adventure, his sunny disposition and "inexhaustible fund of anecdote." At parties, he liked to don his clan's kilts and "skirl merrily on his bag-pipes." He was the best-loved member of the squadron. The Lafayette Escadrille included, at one time or another during its twenty-three months as an independent unit, forty-eight Americans, eleven of them aces—those officially credited with five or more kills.[88]

# Gory Glory, Hallelujah

## THE LAFAYETTE FLYING CORPS

Stung by the loss of several ships to German U-boats and fearful that Europe might be absorbed into the German Empire, the United States on April 6, 1917, declared war on the Central Powers. The nation had only fifty-five military planes—none of them built for combat. Only now did Americans begin to see how far they had fallen behind in the past decade.[1]

Arthur Bluethenthal, a Jewish boy from Wilmington, went to war in early 1916 as an ambulance driver in the ghastly Verdun sector, where he won the Croix de Guerre, and later contended against the Austrians in Serbia. In May 1917, longing to fight, he joined the French Foreign Legion and was soon assigned to flight training. The war was nearing a crisis, with General Erich F. W. Ludendorff striking massive blows at the Allies north of Paris and near the English Channel. An All-Eastern football center at Princeton in 1912 and briefly assistant coach at the University of North Carolina (UNC), Bluethenthal was too big for pursuit planes and so was assigned to bombers. This made him a member of the Lafayette Flying Corps, a designation for all 268 Americans who had served or would serve in French aviation units.[2]

On leave in late 1917, Bluethenthal visited Wilmington, where, a hero of both battlefield and playing field, he was wildly admired.

Dressed in his powder blue, red-capped French uniform, he almost set off a riot at the theater one night, the audience, at his entrance, rising as one to sing the Marseillaise. Back in France, he was assigned to Escadrille 227, the only American in it, and to a single-engine Breguet bomber. On Paris furloughs, comrades knew him as the life of every party and singular in his ability to sleep through air raids and shellings by "Big Bertha" guns seventy miles away.[3]

On June 5 Bluethenthal and his observer flew a mission to direct artillery fire during Ludendorff's latest push near Maignelay, on the Somme River north of Paris. This was the third battle of the Aisne, a German drive to within fifty-five miles of Paris, and it met the first major use of American infantry and strong opposition in the air. Bluethenthal's end was mercifully swift. He and his observer were surprised over the lines by four German planes. The Breguet was hit and both men died in a flaming crash. But the offensive was broken, and the war began to turn decisively in the Allies' favor. Even the sky was now safer since the feared German ace, Manfred von Richtofen, had been killed in April.[4]

The first Wilmingtonian to volunteer and the first to die, Bluethenthal was buried near the crash site, but his body was later exhumed and returned home. His death was honored in an hour-long suspension of business in Wilmington on June 20, 1918, and in a memorial service at the Academy of Music. Princeton, where he had graduated with honors in 1913, established a scholarship in his name; in 1928 Wilmington's municipal airfield was named for him. France posthumously awarded him a second Croix de Guerre.[5]

North Carolina mourned the loss of its third airman to enlist before America entered the war and soon lost the fourth and last. James Henry Baugham was a native of Washington, North Carolina. Like many airmen, he was an athlete, having briefly played halfback at North Carolina College of Agriculture and Mechanic Arts (NCA&M) as a freshman in 1916. He was also a daring race car and speedboat enthusiast.[6] Such youths, physically adept and audacious, inured to speed and quick, competitive decision-making, were ideal for aerial combat.

Baugham quit college in 1916 to enter the civilian Curtiss Aviation School at Newport News, Virginia. There he earned a flying license in 1917 just as America entered the war but, at age seventeen, was rejected by the navy as too young to serve. He was in France by July and soon was accepted for military flight training with the French. In six months of advanced lessons, Baugham earned repute as the most skillful and fearless pilot in the camps where he was billeted, performing loops and vertical spirals with a veteran's aplomb in planes not designed for acrobatics. This was not for show but, as he wrote home, "to be prepared to put my machine in any position I want to at the front and . . . meet any emergency." In December he was assigned to a new unit, Escadrille 157, as a pursuit pilot in a Nieuport biplane.[7]

During the winter Baugham logged many hours in the Vosges Mountains near the Swiss border, learning the risks fliers face even when no one shot at them. While he was on a mission in February 1918, "all hell broke loose." Coming upon Germans pounding "a little village with artillery fire," his plane danced about from explosions below, "plunging every which a way." His motor began to miss, and he turned for home but was only halfway back when the motor died. Landing within French lines, he called his base for a car to pick him up and was soon safely back in camp.[8]

At nineteen, the youngest American to fly for France, he was a war correspondent's darling. Over enemy lines in June, he spied an observation balloon but almost at once saw an enemy scout plane, a more tempting target. Signaling his squadron (which missed the signal) to follow, he gained full speed and began firing within a hundred yards of his prey, apparently crippling the plane. At that moment, however, his motor again died, and he glided toward Allied lines. Searching for level ground, Baugham found "a fairly good place" and landed, stepping out onto the body of a German in a field of corpses, with bullets whizzing by his head. He realized immediately that he was in no-man's-land but had no idea which side was which. Fearing that he might be shot if the Germans took him and found incendiary bullets in his machine gun, he removed its three hundred rounds and hid them under a corpse in an open grave. When he headed back for his fur coat, he found himself caught in a cross fire from both sides.[9]

A German plane overhead drew Baugham's attention to gunfire

aimed at it from his right. He ran that way, but the cross fire became so intense that he "had to fall face down, and I did not move far, for I guess five minutes." Finally, he took out a handkerchief and waved it toward the French. When an officer beckoned from some woods fifty yards away, he ran toward the man, crashing headlong into him. He told a French general that he had seen about three hundred dead Germans but only two French bodies near his plane. "It made him so happy that he gave me a dinner" and praise "for being a good soldier."[10]

An aggressive warrior, Baugham was in many fights. He won the Croix de Guerre with two citations for downing two enemy planes and believed that he had destroyed at least two more. In June he was transferred to Escadrille 98 and a state-of-the-art Spad biplane. At 4:30 P.M. on July 1, 1918, he engaged three Germans near Soissons, fifty-five miles northeast of Paris, near the site of Bluethenthal's fatal crash. Badly hurt and his plane damaged, he crash-landed within French lines. He was rescued but died the next day in a field hospital, posthumously winning the Medaille Militaire. Ironically, his prowess was featured in a 1958 Warner Brothers movie, *The Lafayette Escadrille,* a unit to which he never belonged.[11]

## LES ENFANTS DE GUERRE

Warrenton had never experienced anything like it, maybe never would again. Around 11:00 A.M. on Sunday, October 21, 1917, a biplane appeared in the northeast, circled over town, and landed at the Connell farm, beyond the city limits. It was the first time the town had seen an airplane, as the Great War in Europe, now almost three years long, raged on.[12]

From the cockpit of the Curtiss Jenny climbed Lieutenant Samuel M. Connell, a local lad famous for his flying skill. He had recently been named one of seven airmen to make the first transatlantic flight to Europe, in American-built Handley-Page bombers, in an expected seventeen hours. It would show that planes could reach Europe this way far faster than in crates. (After further study, the army called off the experiment.) Amid a flood of questions, Connell told the crowd that he had flown from Langley Field at Hampton, Virginia, to see his parents. The first Tar Heel to fly home from military service, he promised a show that afternoon.[13]

At two o'clock, reported the *Warren Record*, hundreds thronged the field as the pilot donned flying clothes, mounted his plane, and speeded the propeller. "Bated breath could force no cheer—it was a moment of tense nerves and superb interest." The plane moved. "The spell is broken, under . . . the master control of the pilot [it] leaves the earth," sailing upward. At three hundred feet, "he banks . . . , makes an Immelman [a quick climb into a half turn] and swoops, like some great hawk hungry for prey, back towards the crowd." But he seemed to be too low. "With lightning rapidity he comes, . . . many fall flat, he shoots the juice . . . at twenty five feet up and climbs . . . beautifully" into "graceful turns, before gliding to the earth, kissing it as softly . . . as velvet."[14]

Before leaving on Monday, Connell was in the air again. At three thousand feet, he performed Immelmans, sideslips, spirals, stalls, tailspins, and loops, finally descending to the "cheers of . . . friends and a shower of flowers." As he headed northeast, telegraphers wired back that he was successively over Macon, Vaughan, Littleton, Thelma, Weldon, Gumberry, Seaboard, Branchville, and on into Virginia. It was the least costly excitement that Warren had ever known, its federal taxes brilliantly at work.[15]

Like Connell, who learned to fly in North Carolina's National Guard, other Tar Heels were fliers before entering military service. Greensboro's Forrest Wysong, Raleigh's Djalmar Marshburn, Wendell's Bill Winston, and others had built and flown gliders as teenagers. Brothers Harmon and John Rorison of Wilmington, James Baugham of Washington, and Kenyon Woody of Wilmington were among others with prewar flight training. Gibsonville's Hubert McLean and Raleigh's Seasbrook Renn had flown army planes on the Mexican border. Milburn Bishop as a lad in Wilson had once earned fifty dollars at Belhaven for flying a circus balloon when its owner became ill.[16]

Lieutenant Connell gained proficiency in the army's rigorous training program, where almost all recruits were college men. Among hundreds of wartime Tar Heel fliers, a large portion had attended NCA&M. Scores more, like Connell, came from UNC, others from Trinity (now Duke University), Elon, Wake Forest, Davidson, and other state and out-of-state colleges. Many left school as undergraduates in 1917 and 1918 for air service.[17] In November 1917 seventeen-year-old Delton Cahoon, an air-minded Tyrrell County lad, drove his father's car to Kinston without permission, sold it for $125, and volunteered for the army air service. Police sent him home to an irate father.[18] By and large, the recruits of 1917–18 were "the best and brightest" the state could offer. The air war was thus a middle- and upper-class affair, like unto the Middle Ages, with its own rules of chivalry and ruthlessness.

An army cadet by mid-1917, Samuel Connell spent seven weeks at Georgia Tech's ground school. Twelve-hour-a-day classes included construction, care and rigging of engines and planes, photography, and cross-country and general flying theory. He heard lectures on gas particles, astronomy, meteorology, bombs and bombing, radio, wireless telegraphy, machine guns, aerial reconnaissance, observation, and signaling. But most trainees assumed that their duty, whether in planes or balloons, was to spy on enemy movements and direct artillery fire. Few yet realized that balky motors might be a greater threat at the front line than German planes and artillery. Survivors went on to flying school. Connell, for example, attended the program in Dayton, Ohio. On the first day he sat in a rear cockpit, his trainer in front. He touched the hand and foot controls on the second day, operated them on the third, and sat in the pilot's seat on the fourth. After a half-hour of dual-control flying, he soloed.[19]

The first dual-control flight could be magic. After checking the plane, wrote Warrenton's Donald Daniel, "the Lieutenant and I crawl in. Then the mechanic runs in front" and says "the ignition is off, so we can spin the propeller . . . filling the cylinders with a vaporized mixture of gas, so the engine will start easily." Now came "the exciting moment, the engine was warming up and I was . . . hooking my life belt. . . . Before I knew it we were 'taxiing' . . . faster and faster, until . . . the reaction of the air on the wings" exceeded the plane's weight. "We were going . . . 60 miles an hour when we left the ground and . . . it seemed . . . a dream." The instructor circled and dived to show how ailerons, rudder, and elevators held the plane on its vertical, lateral, and horizontal axes.[20]

"An amateur like me," observed former UNC track star Phil Woollcott (whose mother caused the accident at Raleigh's 1910 air show), "has little time to revel in the beauties of nature." He kept an ear tuned to his motor's hum, his eyes on the clouds, altimeter, barograph, pressure gauge, and tachometer. He worked aileron and elevator controls, "blows his frozen nose . . . [and] twists about watching for other machines and leaning over to get his bearing" from below.[21]

The solo could encompass the most terrifying moments in a trainee's life. Greensboro's Eddie Klingman trained with the Canadian Royal Flying Corps, whose rumored goal was to run the students through fast to kill off misfits. "I was just unconscious with fright," Klingman admitted. "God guided that airplane. . . . I forgot everything I had been told . . . , and when I tried to land I missed the . . . field and had to try it all over again." Some neophytes—including Julius F. Andrews of Durham, Kingsley Culbertson of Cabarrus County, Alexander H. Pickell of Raleigh, Wilder Tomlinson of Louisburg, F. H. Slater of Trenton, and Alfred A. Williams of Greensboro—died from training accidents or illness, many more, in fact, than were killed in combat.[22]

Salisbury's Tom Cushing, a Davidson graduate, worked his way up to instructor in Louisiana but in 1918 suffered a near-fatal crash. He spent the next thirteen years in Walter Reed Hospital, much of the time encased, shoulder to toe, in plaster of paris. Cushing had fifty-seven operations, thirty-three of them major! He hobbled out on crutches in 1931 to marry and then served Asheville as a journalist and secretary of the Retail Coal Merchants Association. Robert Ripley of *Believe It or Not* called him "the gamest man in the world."[23]

Airmen wrote home often—sometimes in clumsy efforts at reassurance. "I'll tell you frankly," David Davis wrote his mother in Red Springs, "we . . . are not expected to last an average of over two weeks" in combat. Winston-Salem's Elbert E. Wilson noted that six men had died in camp the previous week. Wilson was learning to aim to hit the pilot; it's better "to destroy the pilot than the plane."[24]

Despite the frailty of the airplanes, the pilots' inexperience, and the lack of parachutes (not issued until 1919), accidents were rarely fatal. Samuel Connell was circling over Cincinnati when his motor died at four hundred feet and he went into a steep dive. "Then it happened—the miracle. Within a few feet of the ground . . . the lunging plane straightened and shot like an arrow parallel to the earth until it came down, as light as a feather. . . . 'By George,'" he exclaimed, rising from his cockpit, "'it was just luck.'"[25]

Edwin S. Pou, son of Smithfield congressman Edward Pou, was training solo in Canada when he encountered fog and emerged over Niagara Falls. He turned toward the airfield but ran out of gas and was forced to make an emergency landing. Then he saw "a great throng of spectators" at the falls. Noting a small opening nearby, he landed safely, "save a severe shaking up." A witness said that Pou showed an "American spirit of . . . forethought" for others, even at his own peril. Major Claud K. Rhinehart, of Haywood County, en route to Fort Worth in March 1918, saw a wing strut snap at three thousand feet. With his copilot at the stick, he leaned from the cockpit "and by physical force" held the wing in place until they reached the field. His "famous record as a polo player" was cited as instrumental in saving men and machine.[26]

Once at ease among the clouds, an airman might even write letters as he flew. Raleigh's Joseph Boushall, over Arkansas, wrote his family: "You have no idea how wonderful it is way up here with nothing but pure fresh air between you and the ground." He spotted a train, "so think I will give it a race. . . . I see a town" and went down to "see if the people are up yet, as it is rather early. Here I go down and, yes, the people are up." Connell, at Houston, described a spiritual grandeur of open-cockpit flight that turned farmers into poets. "The artistic brush could not portray" the morning sky, "the ocean in all of its . . . silent grandeur, falls far short. Beneath me the clouds rolled like billows, and in their somber impressiveness, played upon the sunlight," making "a scene indescribable for its beauty. . . . I actually was persuaded . . . that I might see an angel floating along with me." Thus it was "a revelation which I shall try never to forget."[27]

In France, pilots were now in critical need and training was conducted at a frenzied pace, often in foul conditions. "The sun is a luxury here," wrote Fairmont's Edgar E. Cobb from Issoudun, the world's largest airfield. "We wear hip boots and half the time these are muddy above the knees. . . . I have not tasted a thing now for three days except hot chocolate" at the YMCA. His instructors were French, so he did "all the motioning, jabbering and gestures known to man." Six hours of flying, estimated Warrenton-native William W. Palmer, "is equal to about eight hours of plowing."[28]

Combat training was given to only the best cadets and was more feverish than anything Cobb had seen. "Eight or ten hours of 'solo' flying," he said, "then the student is ready for his altitude flights." He had to climb to "2,000 meters and maintain it for an hour." He made a cross-country flight to "3 or 4 cities," landing at each to report back by phone. After twenty-five hours of flying, he earned his pilot's license and knew "all there is to know about flying."[29]

Crashes were frequent at Issoudun, observed Paul Montague of Winston-Salem; funerals averaged one a day. Few attended. "It isn't that we were hard. . . . We held his sacrifice sacred, but put him out of our minds."[30] They lived, if they could, a day at a time.

## DANTE'S INFERNO

In mid-1918 few American fliers had seen combat, but the war's momentum began changing in May as the first U.S. squadrons reached the front. Among them were scores of Tar Heel airmen fresh from training and, for the most part, impatient to fight. "We have only one idea," wrote Raleigh's Djalmar Marshburn to his mother, "to run the Germans out of sky. . . . We are going to surpass the British flyers and you know what a reputation they have."[31] But even those with advanced work in American planes found the French ones too hot to handle and needed further training to fly them.

In July North Carolina newspapers began an almost daily drumroll of aerial deaths and injuries. On June 29 Raleigh's Dudley Robbins and Asheville's John C. Wilford, of the Eighty-ninth Aero Squadron, were bringing their plane home from an infantry liaison when motor trouble forced them down, but they were soon in the air again. A mile west of Chatillon, on the Marne, the plane spun and slammed into the ground, killing them both. "The gasoline tank," wrote an officer, burst and "the aeroplane was almost totally consumed before help arrived."[32]

Observer Lieutenant Jack Lumsden, of Raleigh, age forty-two, and his forty-five-year-old pilot were considered the oldest flying team in the war. On July 24 Lumsden and Lieutenant John C. Miller of Fairview flew in Twelfth Squadron planes on a photo mission near bloody Chateau Thierry, eighteen miles west of Chatillon. Miller had just finished taking photographs behind enemy lines when attacked by six German aircraft, Lumsden by five more. Enemy planes "surrounded Miller, pouring in lead from all sides." Badly wounded by an explosive bullet in his chest and an arm, sliding in and out of consciousness, he wrestled his plane homeward, slumping forward as he was hit again. "By some miracle" he revived, pulled back the stick, and landed. Miller's observer leaped out and saw him "climbing out of the cockpit, bleeding profusely, his face ghastly white." Gasping that they had made a good set of pictures, Miller fell forward and died. He won the Croix de Guerre with palm for two confirmed kills.[33]

Lumsden, a gunner, emerged from the fight unhurt and was on leave in Paris a short time later when his pilot was killed. He himself died on August 16. His pilot on that occasion wrote that they had been protecting an observation plane when attacked by a swarm of German aircraft. Three times Lumsden stopped firing at enemy planes diving at him on one side "to swing his gun to the other side" at those attacking the observation plane. This left him entirely exposed to enemy fire, and on the third such maneuver he took a bullet in the temple. "Even under murderous fire," he knew that he was there "to protect that other machine."[34]

Other Tar Heels died in the war's ferocious last three months. Among them, Jim Sykes of Charlotte was killed crashing behind enemy lines; Elizabeth City attorney Sidney W. White, a former Wake Forest trackman, when his plane caught fire returning from patrol; and Hickory's Lawrence B. Loughran in a dogfight in the Calais sector. Some airmen, like C. C. Olive of Greensboro, died abroad of illness.[35]

Yet most North Carolinians lived through it, a few even in com-

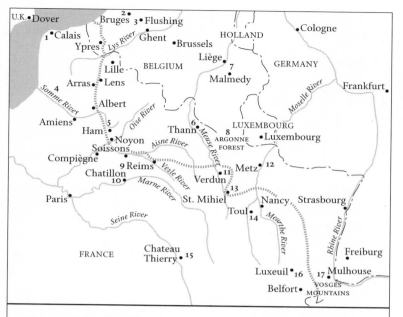

1. Lawrence B. Loughran killed.
2. Don Rion Harris in bombing raid.
3. Harris interned after his capture.
4. Erich F. W. Ludendorff's offensive, May 1917, third battle of the Aisne.
5. Site of James R. McConnell's fatal dogfight, March 16, 1917.
6. Kiffin Rockwell killed.
7. Harmon Rorison downs enemy planes.
8. Site of Robert Opie Lindsay's most daring feat.
9. James Henry Baugham killed.
10. Dudley Robbins and John C. Wilford killed.
11. Site of war's heaviest fighting, in which the Lafayette Escadrille was involved.
12. Site of James Allen Higgs Jr.'s balloon exploits, July–October 1918.
13. Lindsay downs German planes.
14. Site of Paul Montague's force-down, July 1918.
15. John C. Miller killed.
16. Training camp of Lafayette Escadrille.
17. Rockwell becomes first American to shoot down an enemy plane, May 18, 1916.

fortable inactivity. Don Rion Harris, of Buncombe County, styled by the *Charlotte Observer* as "one of the leading society men of the city," joined the Royal Air Force (RAF) in 1918 and flew bombing missions over Belgium and France. On August 16, in a raid on the docks at Bruges, his plane was hit by "archie" (antiaircraft fire), which pierced the motor and gas tanks. He still managed to drop his bombs on the docks and even take photographs of the Bruges-Zeebrugge

Canal. "We fired our distress signals," he wrote home, turned toward Zeeland, twenty miles north, in neutral Holland, and landed "in an ignominious potato patch." Dutch military police took Harris and his observer to a nearby town, whose mayor treated them to wine and cigars before sending them on the next day to authorities in the town of Flushing.[36]

Here Harris found English widely spoken and that evening was allowed to stroll the beach boulevard like any tourist. Interned at The Hague for the duration of the war, he was appointed to the staff of the American military attaché, rented a small apartment, and sent home for such things as books, pajamas, cigarettes, and toiletries. He was "the only American flyer ever interned in Holland," he wrote, so "a man of importance. Everyone wants to see me and I am . . . an object of curiosity."[37] The war was over in three weeks.

A force-down behind enemy lines was another matter. Paul Montague of Winston-Salem believed that he might have downed as many as a dozen German planes, but in the heat of combat it was often impossible to say who got the ones that fell. On July 31, 1918, Montague, of the Ninety-fifth Squadron, which included Teddy Roosevelt's son Quentin, flew with others in the Toul sector as cover for a reconnaissance mission over Germany. Coming upon a dogfight, his motor was riddled with bullets and he landed in German territory.[38]

Montague tumbled out "and ran to the nearest ditch." "Heilt [Halt]," demanded a German, "and a half dozen rifles covered me. . . . At 9 o'clock that night I was allowed to stop marching for the first time. I had been given no food. When food was brought it consisted of black bread, which I have been told had sawdust in it, and a sour black soup." Montague was sick for two days, moving from one prison to another. At length he reached a "reprisal camp" in downtown Karlsrue, so situated to deter Allied bombing of the city. "It didn't work," he adds, but the war soon ended anyhow.[39]

Force-downs on friendly soil could ignite impromptu celebrations. One August morning a bad magneto put Wilson's Milburn Bishop down at a French town. The people turned out, the mayor took him to lunch, and he "was treated like the king of some country." That Bishop "owned the town would be putting it altogether too mildly." Mechanics he called fixed the plane, and he did some acrobatics. On landing, "they just ran up . . . and everyone clapped . . . and shook hands with me." As he was leaving, a pretty girl gave him a bouquet of flowers (one of fifteen he received) and got his address. No sooner did he reach his base than a letter arrived to "my Airman Prince." His mates begged piteously though unsuccessfully for the town's name.[40]

With combat came hardened attitudes and fading illusions new to many fliers. Reidsville's John Y. Stokes, a gunner, wrote home in October 1918 that sometimes "returning . . . from a raid, we sight a road filled with Boche troops and . . . turn our . . . guns on them. That's great sport, but if your motor stops . . . you're out of luck." Besides "killing a good many . . . that kind of treatment lowers their morale and makes them glad to be taken prisoners."[41]

Stokes's view resulted from the fast pace of his work and propaganda about barbarian "Huns." Usually up by 4:00 A.M. and off at first light, he flew three missions a day, one formation landing as another left. "This morning . . . down out of the clouds," he wrote, "poured what . . . seemed to be about 50 Boche. . . . They opened up fire and so did we, but being outnumbered we dived into a cloud to lose them. . . . When we came out . . . five of them still stuck with us . . . and all of us together got two," the others fleeing. Over Metz, his formation dropped "36 bombs, about as large as a man," all in the center of town.[42]

On another run, Stokes's formation was broken up by mechanical problems, but his pilot continued on and joined another formation. Hit by "archie," his plane was briefly out of control, but the pilot, despite a badly missing motor, reached the destination and dropped his bombs. "By this time," reads Stokes's Medal of Honor citation, their motor "had died completely. They were then attacked by an enemy combat plane, but, by diving with a dead propeller and with continuous fire from Stokes's gun, the attacker was held off until they crashed unhurt in treetops within Allied lines." Stokes had four confirmed kills; several others, behind enemy lines, could not be corroborated.[43]

Officially, the most successful Tar Heel airman in World War I was Lieutenant Robert Opie Lindsay, of Madison, in Rockingham

County, formerly a basketball player at NCA&M. After his early training was bolstered by two months of acrobatics at Issoudun and gunnery at Casaux, he was assigned in August 1918 to a 139th Aero Squadron Spad on the St. Mihiel front. Lindsay was in twenty fights in ten weeks; he survived two four-mile dives in bullet-riddled planes. His career was capped when he and another pilot, "using cowboy tactics," downed two Germans from a formation. Chasing two more fleeing homeward, the Americans downed them over their own German field, a feat the *Washington Post* called "the most daring mission ever witnessed on the Western Front."[44]

Shifted to the Argonne sector, Lindsay's was one of three planes near Bantheville on October 27, 1918, that attacked three Fokkers. He forced one down and was chasing the others when eight more appeared. He "flew straight through their formation, gained the advantage, and brought down another," the first of his four on this front. With six kills, he was the lone Tar Heel in an inflated list of eighty-two American aces and won the Distinguished Flying Cross.[45]

Wilmington's Harmon and John Rorison took flying lessons on Long Island before enlisting in the Royal Air Force in 1917. On July 10, 1918, Harmon and another pilot engaged a Pfalz scout-plane and "shot it to pieces." The next day, he got three Fokkers. On the thirty-first he downed an Albatross, in August and September two more. He was credited with only one of his seven victories, the rest going to fellow RAF airmen. He soon transferred to an American squadron.[46]

In the last week of the war, Harmon Rorison was on a bombing run with four other airmen south of Malmedy when attacked by eighteen Fokkers. Three American planes fell, and a bullet hit a bomb he was carrying but did not set it off. Rorison continued to fight and downed two of the Fokkers. Wounded, with gas spraying from his emergency tank, he downed another enemy before one of his guns was hit. He then had to fly through heavy ground fire as he returned over the lines to his field. His brother John won the Distinguished Service Cross (DSC) for heroism over Beaumont on November 3.[47]

Navy pilots experienced dangerous, though usually less challenging, tests. Lieutenant Edwin Pou, flying a seaplane on the southwest-

Lieutenant Robert Opie Lindsay of Rockingham County, whose six kills made him North Carolina's only World War I ace. He is shown here as a senior at present-day North Carolina State University in 1916. North Carolina State University Archives and History, Raleigh.

*Gory Glory, Hallelujah*

233

ern coast of France on September 9, 1918, depth-charged and destroyed a German mine, then another on October 26. Two days later, landing at his base, Ile Tudy, his plane hit a spar-buoy and burst into flames, killing him and two crewmen. He was awarded a posthumous Croix de Guerre. Forrest Wysong saw flying-boat service at Dunkirk and elsewhere.[48]

Not all airmen flew planes. Lieutenant James Allen Higgs Jr. of Raleigh won the DSC in a balloon. When attacked by planes, balloonists, who carried parachutes, could only jump and pray. Higgs was honored for action on July 31, August 21, and October 29, 1918, south of Metz, while under attack by enemy planes. On each occasion, he aided a partner out before jumping himself. In the October 29 instance, while he was directing artillery, his balloon caught fire. He waited to jump "until he had assisted his companion in escape." Each time, he "at once reascended in a new balloon."[49]

Tar Heel airmen winning combat medals also included Warrenton native William Palmer, an ace whose home was in South Carolina, and Tarboro's Spencer Hart, among others.[50] Palmer flew with Eddie Rickenbacker's famed Ninety-fourth Pursuit Squadron. North Carolina lost at least sixteen sons in air combat, most of them in the last four months of the war. Tar Heels appear to have downed several times their number of German fliers.

In 1917 and 1918 Tar Heels at home might sometimes glimpse a bi-plane winging between Virginia and South Carolina airfields, even a refueling or breakdown that caused it to land briefly. Occasionally, a seaplane passed along the coast. But far fewer Tar Heel civilians saw an airplane close up during the war than in 1911. Aviation ceased to be a palpable presence until late in the war, when the state finally acquired a few aviation facilities from the burgeoning growth of airplane production and pilot training.

If anyone could talk North Carolina into embracing airplane production it was Harry Atwood. In the course of six years' flying, Atwood had set long-distance records, had won air derbies, and was the first pilot to land a plane on the White House lawn, as well as the first to fly a plane over New York City. He was the embodiment of the reckless pioneer who would risk everything for a chance to do something new and flamboyant.[51]

In February 1918 Atwood showed up in Raleigh with an intriguing idea. Planes needed wood, and North Carolina had plenty. They needed textiles, and this was a leading cotton state. They needed mild flying weather, and the state's temperate climate extended through ten months of the year. These blessings could help answer acute national needs. No one, he reminded Raleighites, knew how long the war might last. The government wanted 100,000 planes to help ensure Allied victory; it would need a million propeller blades, a quarter-million motors. After the war, as many as fifty airmail routes would be established, and Raleigh was already on one of those projected by the Wright Company. Why shouldn't this state, where aviation began, do its part right here in Raleigh? Why wait another day to start? On February 28 Atwood gave a stem-winder before a gathering at the Woman's Club auditorium, and on March 17 the chamber of commerce voted to investigate Raleigh's potential as a center for airplane manufacture.[52]

With the permission of the Curtiss Aviation Company, Atwood meant to build Curtiss F-boat seaplanes, but incorporating his own innovation, namely, a veneered plywood fuselage and wings. Such a plane would have several advantages over the standard F-boat. Be-

# What We May Expect in the Near Future.

**Cartoonist Billy Borne anticipates airmail in 1910. From** *Asheville Citizen,* **September 28, 1910.**

sides stronger wings, it would not soak up water, whereas existing machines absorbed up to two hundred pounds and sometimes became waterlogged. It would be faster than regular F-boats and need a smaller tail assembly. It would, in fact, be the best flying boat ever built.

The chamber of commerce sent a delegation north with Atwood to learn more. He took the members to Wilmington, Delaware, where he had run a flying school for DuPont and built some of its training planes. In New York, he led them to the Aero Club of America, which endorsed his plans. "Aeroplanes Soon to Be Flying over Raleigh," heralded John Asley Park's *Raleigh Times* on March 27. The new Carolina Aviation Company's offices opened in the Commercial Bank Building. J. Melville Broughton (a future governor) was secretary and Park treasurer.[53]

The company assured the U.S. government that it could build

planes for half of what it was paying other manufacturers and deliver up to five hundred planes in the first six months of production. There was talk of a $5 million navy contract. Atwood was working in Raleigh on a prototype to be ready in sixty days. The company would hire two thousand people and might, in time, build hangars, parts, "aero-cycle cars," lighter-than-air craft, and so on. It would open a flying school; carry mail, freight, and passengers; and "conduct experiments." By June 21 the first plane was said to be assembled and to have satisfied all tests.[54]

But the project was mostly a matter of bells and bugle calls. Had the chamber of commerce looked more closely, it might have found at Sandusky, Ohio, and Williamsport, Pennsylvania, among other places, people who knew Atwood only too well. In those towns, he had promoted schemes to build seaplanes or a radical new type of motor. The projects collapsed, and investors lost a lot of money. In some quarters Atwood was thus becoming known as an unrelenting con artist, and Raleigh knew nothing of it.[55]

There is no evidence that the prototype seaplane was actually built. Moreover, the navy showed no interest in Atwood's ideas, though it yielded to pressure from the Senate Finance Committee, headed by Furnifold M. Simmons of New Bern, and placed two small orders with Atwood. He encouraged a bidding war for his plants in Fayetteville, Durham, and Goldsboro, as well as Raleigh. In the late summer Raleigh wisely opted out, and Atwood selected Smithfield and Goldsboro, which already had veneer mills. Because the project was being bankrolled by federal money, Smithfield and Goldsboro had no problems with Atwood's financial arrangements. His Cadillac was often seen roaring over rough roads in Wayne and Johnston Counties; he could cover the forty miles between Smithfield and Raleigh in a record forty-five minutes.[56]

By the fall the navy had an F-boat base at Morehead City, and Atwood promised to have his first seaplane there by October 15. But when the truck transporting the plane finally reached the coast in December, the war was over. (Nationally, Washington had already canceled orders for 13,000 planes.) When Atwood's seaplane made its first short, low-level flight in January, a navy inspector found it to be 470 pounds overweight and altered without Curtiss's permission.

Atwood protested that the aircraft, with its OXX-6 motor and 40-gallon fuel tank, had a range of 325 miles and could fly at 4,500 feet, both spectacular attainments for a seaplane. After the inspector left, the builder added that it flew well into a 40-mile wind.[57]

The navy demanded changes, which he made reluctantly. But the plane was still too heavy and, when tested, flew only about a mile at an altitude of twenty-five feet. Maybe a 150-horsepower Hispano-Suiza motor, the inspector felt, would improve it. The navy agreed to buy four F-boats at $16,000 each. The Carolina Aviation Company built them at Smithfield and Goldsboro in 1919, but all were rejected after trials at Coconut Grove, Florida. By then, Atwood was in New York plotting new schemes and leaving more battered bank accounts in his wake.[58]

Morehead City's seaplane base did not, however, depend on Atwood's F-boats. Expecting trouble from German U-boats, navy secretary Josephus Daniels in December 1915 had sent his undersecretary, Franklin D. Roosevelt, to North Carolina to find a suitable site for a navy air training camp. The *Raleigh News and Observer* noted Roosevelt's presence in Carteret County and a rumor that the navy wanted to locate the training facility at Camp Glenn, a Morehead City National Guard camp named for former Governor John B. Glenn. In 1916 Congress voted funds for construction there, but higher priorities held the base up for two years.[59]

Harry Atwood's seaplane in trials off the North Carolina coast, 1918. Courtesy of Peter M. Bowers.

With U-boat activity rising on the Carolina coast, the need for a patrol base was clear; by August 1918 the camp was under construction. At first it was to house a small detachment from the naval air station at Hampton Roads—one plane and eight men—but was soon upgraded to a permanent facility. Governor Thomas Walter Bickett tendered the site to the navy in late August with the understanding that $2 million in federal money would be spent there. The navy objected to the stench of a nearby fish oil factory but was assured that the problem would be remedied. Twelve Curtiss planes and 320 officers and men were stationed at Camp Glenn to patrol the coast as far south as Charleston. The first seaplane arrived in early September, just after the *Diamond Shoals Lightship* and two other vessels were torpedoed and sunk off Cape Hatteras. It was followed by bomb-carrying patrol planes, equipped with carrier pigeons to relay U-boat movements and the location of sinking ships. But no enemy activity was encountered in these last days of the war.[60]

Camp Glenn fliers now turned their machine guns on the local duck population. The assault, launched before duck season began on November 1, drew little criticism since millions of them flocked to these waters each fall. Moreover, duck was a welcome supplement to the navy diet, which depended to a large extent on fish. (Aerial hunting violated the state's only aviation law, which prohibited shooting wildfowl from flying machines.)[61]

At the height of activity at Camp Glenn, the navy was said to have up to five hundred men and fifty planes there. When it was decided to end its operations in early 1920, the Coast Guard took over the facility for experimental use of Curtiss HS-2L seaplanes for coastal patrol. Their prospective functions included finding distressed ships, helping to save crews, spotting rumrunners, and reporting fires. The Coast Guard began to operate from the camp in late March, anticipating that it would have about one hundred personnel.[62]

An unexpected use of the seaplanes was spotting fish for the local menhaden business, Carteret County's main industry. In the early morning Coast Guard pilots flew out at about five hundred feet, and, on seeing large schools, returned to report them. Although the menhaden season usually began in October, fishermen found that they could now make catches as early as July. The base was also a refueling station for Norfolk-Charleston air traffic and other routes. Despite some successful search-and-rescue and law enforcement work, proving the peacetime utility of Coast Guard planes, the seaplane base closed in mid-1921, after less than three years of operation.[63]

Camp Glenn was also the scene of a successful seaplane experiment. In 1922 Lieutenant Commander C. C. Van Paulsen began work there on a method to save ships and crews in distress. It involved a plane carrying out a line from a position on shore to a foundering ship as far as a mile or more at sea, whereas Coast Guard cannon could shoot lines no farther than about seven hundred feet. The procedure was adopted at some Coast Guard stations, which also used the planes to take food, fuel, water, and medication to distressed crews. The procedure reportedly helped save many lives before more modern methods replaced it.[64]

The only other military aviation facility in the state during World War I was at Charlotte's Camp Greene. The base was built in 1918 for aviation repair, communications, artillery training, and an ordnance depot. The *Charlotte Observer* noted on June 1, 1918, that four planes had been assembled at Camp Greene for study by cadets and mechanics and that others were expected. It does not appear, however, that more aircraft were delivered or that any flying was done there; the camp was dismantled as soon as the war ended.


## THE AMITYVILLE HORROR

As the war wound down, the mad scientist, Dr. William W. Christmas, was perfecting plans to give America the pursuit plane it needed to bring the fighting to a quick conclusion. The mighty Battle-Cruiser he had tried to palm off on the government in 1915 had vanished in the mist, but in 1917 he set out to build the world's fastest plane, the amazing Bullet.[65]

In 1917 Vincent Burnelli, a brilliant engineer, was working in the Continental Aircraft Company plant at Roosevelt Field, Long Island. One day, Christmas and some of his cohorts visited the plant. He told Burnelli and other company personnel that he was buying Continental in order to build a super pursuit plane. Its frame would be of hickory, not spruce, and strong sawmill steel instead of the malleable steel in common use. Having spent years studying bird flight, he wanted his plane to have flexible upper wings that could be altered in flight from a drooping contour to straight. As speed was gained, the rear part of the wings would float upward rather than fighting the air and disturbing the plane's balance. The wings would also be cantilevered; no struts or wires to create drag would mean a faster, yet safer machine.[66]

Drawing the group close, Christmas disclosed in a low voice that his planes would be used to kidnap Kaiser Wilhelm. "'They'll be flying around the castle, shooting everything that moves. It will be blocked passage in or out.'" The rest of the plan was secret, but he confided that "'the . . . Bullet can fly *over* it, with a passenger on board. . . . And when we get the ole Kaiser . . . , he'll call off the war or else.'"[67]

For his laudable purpose, Christmas had several backers, headed by the McCorey brothers, Alfred and Harry, who were Hartford, Connecticut, bankers. Alfred seems to have been the chief sponsor, investing a reported $400,000. The inventor himself had ties with General William L. Kenly, chief of air services for the Signal Corps, and retired Admiral William Crumshied. He was also fortunate in his wife (or mistress)—"a luscious Latin beauty named Maria" who "thought he was wonderful." This group had in common an admiration for Christmas and an eccentric conception of airplane design.[68]

The reborn Cantilever Aero Company began building the wonder plane. Christmas used no blueprints in his work; at intervals he dashed off sketches of various parts for the engineers to fabricate. But Burnelli soon concluded that the inventor had little idea of what he was talking about. If Christmas asked a worker what he was doing, it was enough to reply that he was "making the fleebus for the repple." Yet the Bullet took shape as an unrivaled warplane. The upper wing was flat across the fuselage, so the pilot could fire his gun over it, with a full view front and back; the lower wing was staggered back for a clear view below. Christmas used his connections to obtain a 175-horsepower Liberty motor, giving his plane exceptional power; other features ensured high maneuverability. Word got out and interest grew in aviation circles. Maybe America would put a combat plane, a marvelous one at that, in the war after all.[69]

Eventually, however, Burnelli told Christmas that the upper wing must not be built as designed. The inventor had insisted that it be in two sections, its steel parts welded together. This left a joint that "you could have snapped . . . over your knee," declared Burnelli. But Christmas persisted; he covered the wings with a burlaplike fabric that stretched where it should shrink and gave it numerous coats of acetate dope. According to Burnelli, the wings were so heavy that they had to be winched into place, and the tail assembly was too small for so heavy a plane. Burnelli resigned from Continental Aircraft over the issue.[70]

Visitors saw a sleek plane with a deep, short fuselage "of singularly clean lines, free of struts and wires," painted "a dazzling white," with red trim. One observer called it "as trim and speedy-looking a craft as ever graced a flying field." It was expected to fly over two hundred miles an hour, an amazing velocity for the time.[71]

The war had just ended when veteran Cuthbert Mills gave the Bullet its first test flights, evidently on December 3 and 7, 1918 (though the dates are debated). J. D. Van Vliet, a Christmas admirer, reported that, on the first of five flights, it "took off after a very short run and showed a speed of 190 m.p.h.," despite weather that grounded other planes on the field. Mills praised its lateral stability and rate of climb as "seemingly out of proportion to the power available." It turned and steered easily, almost flying itself. A witness, Lieutenant E. E. Hall, heard Mills say that it "was the fastest he had ever flown . . . the performance . . . was wonderful."[72]

According to Van Vliet, the Bullet attracted wide attention, "especially in France and Germany," where flexible wings had long been under serious study. But much, if not all, of his information evidently came from Christmas himself. Vincent Burnelli, who stayed in touch with the project, gave a very different account. Several pilots, he asserted, refused to fly the Bullet before Mills agreed to do so. On his first try, the wings fell off and it crashed, killing him. "At least, I figured," wrote Burnelli, that this would be the end of the Bullet. "But I underestimated the Doc." It failed, Christmas insisted, be-

The Bullet, 1918, was intended to be America's first pursuit plane. Its designer, Dr. William W. Christmas, claimed it could fly over two hundred miles an hour. From *Popular Mechanics*, August 1929, 237. © The Hearst Corporation. All rights reserved.

cause of "pro-German saboteurs" (implying a prearmistice rather than a December flight). A second plane was built, and pilot Allington Jolly was hired "at a huge salary." On his first flight, at Amityville, New York, Jolly flew the craft a short distance but could not control it. A barn "loomed up in front of him," and he "couldn't maneuver . . . over it or around it." He, too, was killed.[73]

Long Island newspapers indicate that Mills's crash occurred on December 30, 1918, from either engine failure or a collapsed lower wing. A third Bullet, after its display at the New York Aeronautical Exhibition in March 1919, crashed because its wings failed during a May 1 test flight.[74] So ended the brief career of the Amityville Horror.

Whatever the number and dates of the flights, Christmas emerged from the fiasco in fine shape, claiming that the Bullet had set a world record of 222 miles per hour. He also enriched himself at his backers' expense and, Burnelli maintained, had "designed a plane so terrifyingly powerful that the Germans had had to sabotage it twice." Like Harry Atwood, William Christmas had more tricks up his sleeve in the years ahead.[75]

. . .

Less than fifteen years after the first airplane flight, hundreds of North Carolinians had become expert pilots. Some of them would make a career of flying, others would lend their solid support to building airfields and promoting civil aviation in the state. As before, aviation's progress had demanded a high price in crippling injuries and young lives lost. But the state and country emerged from the war at the end of 1918 with a familiarity and a confidence in aviation technology that would have been unthinkable even two years earlier. That the public would push for the quick adaptation of airplanes to peacetime uses was a foregone conclusion.

# The Greatest Pilot on Earth

At the war's end, the newly named Air Corps was about to build its first North Carolina airfield, part of a new military reservation near Fayetteville named Camp (now Fort) Bragg for Confederate General Braxton Bragg. Its first casualties came on January 7, 1919, when Lieutenant Harley H. Pope and his mechanic, Sergeant W. W. Flemming, crashed into the Cape Fear River near Fayetteville en route to the pine forest site to prepare an airstrip. In March the field was named in Pope's honor. Pope Field in February became home base for the 32d Balloon Company, 84th Photographic Section, and, two months later, 276th Aero Squadron.[1]

Almost immediately the camp made its presence felt in much of North Carolina. In May, Pope airplanes flew to Charlotte to take pictures of possible landing sites and map routes for an "airplane guide book." In June, a Pope airman, flying low because of a defective compass, discovered three moonshiners near Wilmington. Others were soon reporting house and forest fires and helping towns such as Wilmington lay off emergency airfields. Alert Pope fliers were credited with saving the Cumberland County hamlet of Manchester from destruction by flames. When public events were planned at the field, pilots visited towns in the region to drop handbills with the schedule of events.[2]

Pope airplanes were very active in the summer of 1919, performing stunts for bond sales and recruitment drives in towns around the state. Pilots sometimes delivered mail to artillery units in the field. In 1921 a Pope pilot made a transcontinental flight but was killed in a crash as he returned. Meanwhile, Camp Bragg, an artillery base, drew thousands of troops to the area, a blessing to the local economy. The new airfield quickly became an important stopover for north-south military air traffic, but it was some time before a considerable number of planes were permanently stationed there.[3]

For some wartime aviators, the November 1918 armistice brought neither peace nor a quick return home. Paul Rockwell, a correspondent after he recovered from injury in the trenches, learned to fly and fought for France after the war. In 1925 he organized a squadron of Americans to help subdue a rising of the Riffs in French-controlled Morocco, the French themselves being barred by the League of Nations from intervening. Abd-el-Krim and his Berber Riffs were conquered in 1926.[4]

Wilmington's Harmon Rorison, following outstanding work on the western front in 1918, enlisted a year later in the Kosciuszko Squadron of Americans in the Polish Air Force as Poland clung to its newly won independence against invading Russian Bolsheviks. Described as "a five-foot, five inch . . . bantam rooster" of a man, known to his comrades as "Little Rory," Rorison was a welcome addition to his unit and early proved his worth.[5]

On March 5, 1920, Rorison, in a Balilla biplane, was the first person in the squadron to attack Russian forces, hence the first American flier to see action against Communists. Observing a "big concentration of Bolshevik troops and three armored trains" in East Galicia, he made bombing and strafing runs that were said to have "given the Communists hell." But, as he was "not exactly a loquacious man," not much is known of the details of this or his other exploits. Rorison engaged in bombing, strafing, and reconnaissance missions at low altitudes, striking troop trains, cavalry, and machine-gun nests. Once, with broken ribs, he found his way back to Polish lines after being shot down deep in enemy territory. With Russia's defeat, a grateful Poland awarded him the Virtuti Militari, V Class,

Harmon Rorison of Wilmington, shown in Poland in the cockpit of his Balilla biplane. From Robert F. Karolevitz and Ross S. Fenn, *Flight of Eagles: The Story of the American Kosciuszko Squadron in the Polish-Russian War, 1919–1920* (1974), 129, Brevet Press, Inc., and Shrewsbury Collection.

its highest military honor, as well as the Liberation de Pologne and Haller medals.[6]

Some pilots stayed in the Air Corps, including several who gained high distinction. Marine Lieutenant Andrew Crinkley of Raleigh (a Warrenton native) was the first to fly an airboat round-trip across America. In 1926 he set a world record for hours aloft in a seaplane, retiring in 1941 with the rank of commander. Samuel M. Connell rose steadily through the ranks to brigadier general in World War II. Mount Airy's Caleb Vance Haynes, a descendant of Siamese twins Chang and Eng Bunker, made the first American flight to Puerto Rico and first army round-trips across both the North and South Atlantic.[7]

Smithfield veteran R. A. Wellons was still in the army when he made headlines, on December 6, 1922, by flying under all four St. Louis bridges spanning the Mississippi—a first. William A. "Bill" Winston of Wendell was an army instructor in Texas, when, in 1923, he gave a young Minnesotan, Charles A. Lindbergh, his first army flying lessons. Winston was later the top transoceanic pilot for Pan-American and a favorite with passengers for his parlor tricks and laid-back good humor. North Wilkesboro's Chalmers G. Hall, a civilian veteran of the Spanish-American War, became chief of Air Service Procurement in 1919.[8]

Some veterans bought surplus Jennies at $150 or less and stayed in the flying business. Hubert McLean flew home to Gibsonville in one in 1918 and laid off an airfield in his backyard. (McLean Brothers Airport is now the oldest such operating facility in the state.) Some pilots became barnstormers for a public anxious to see the planes and tactics spawned by war. Wilson's J. J. Privette and Milburn Bishop formed an aerial company and bought a Jenny for exhibitions and lessons. Henry M. Westall, formerly with the Canadian Royal Air Force, and Gus Leazar of Morrisville became barnstormers and commercial pilots. Westall, hired by Asheville leaders to promote

**Lieutenant Augustus "Gus" Leazar of Morrisville, a top training pilot in World War I. Courtesy of Hardy Mills.**

the city by air, enjoyed a long civilian career. Wilmington veteran Kenyon Woody and young Warren K. Pennington bought a Jenny and barnstormed from Maryland to Florida.[9]

The gifted Statesville engineer, Raymond V. Allison, who failed in 1910 to become the first North Carolina aerial exhibitionist, joined the army air service in early 1918 and began training at Austin, Texas. He was soon dismissed because, at thirty-one, he was found to be over the service age limit. At the instigation of Senator Lee S. Overman of Salisbury, however, he was training again in a few weeks.[10]

Allison was known for his "picturesque personality, his Van Dyke beard, and other characteristics setting him off distinctively," and he enjoyed "a wide circle of friends." He made second lieutenant, graduated in August from the Texas School of Military Aeronautics, and was slated in October to go to France as a balloon observer. The order was canceled by the war's sudden end, but he was promoted to captain, discharged in May 1919, and joined the reserves.[11]

As a cadet, Allison accumulated numerous flying hours, but his life in and out of the army was marked by one bizarre episode after another. Soon after his discharge, his brother William wrote to him regarding Allison's claim that he had set a world altitude record of

**Warren K. Pennington of Wilmington. Courtesy of Anna Pennington.**

3,800 feet at Fort Sill, Oklahoma. The letter contained a clipping about a Frenchman recently setting a record of 3,100 feet; William urged Raymond to publicize his own feat and "make the Frenchman's exploit look insignificant." Allison gamely replied that he had made his flight against orders and must keep it secret.[12]

According to family legend, Raymond Allison, though not sent overseas, wished his mother to think he was in European combat. He wrote letters "from the front" that his brother William dutifully forwarded home from Chicago, thus bearing no postmark. In one, he told of being forced down behind enemy lines. He was captured by German troops, who wanted to move his plane to a field in the rear, but none of them could fly. An officer went with him to direct him, but Raymond, once airborne, flipped his plane over, ejected the officer, and flew home.[13]

Allison's return to Statesville after his discharge was no less bizarre. On April 16, 1922, police officers went to his home to arrest a domestic servant for selling alcohol. Allison, with a cocked pistol, ordered them to leave. One of them, whom he physically attacked, drew his own gun. In an exchange of shots, Allison was hit in the chest. He ran into the house, returned with another gun, and collapsed in the yard. He recovered and seems to have served no jail time. Soon afterward he moved to Texas, where he had friends, settling at Cleburne. There, in September 1926, he was embroiled in an altercation when he demanded that the YMCA reimburse him for articles stolen from his room. In the course of the dispute, he was shot by an employee and died instantly. Friends cited numerous instances of his kindheartedness, tact, and courtesy. In any event, Statesville would not see his like again.[14]

Many North Carolina military pilots went on to notable careers in other fields. Navy cadet Harold D. Cooley of Smithfield served thirty-one years in Congress, and Claude L. Currie of Montgomery County was dean of the North Carolina State Senate when he retired in 1972, at age eighty. Durham's William D. Carmichael became a longtime comptroller at the University of North Carolina at Chapel Hill; Charlotte's John L. Morehead was a textile magnate, and Raleigh's William H. Stephenson developed the first automatic resuscitator.[15]

The most accomplished fliers to emerge unscathed from the war included America's top ace, Ohio's Eddie Rickenbacker, who shot down twenty-six enemy planes in his short career, and Elliott Springs, of South Carolina, who downed twelve. Perhaps an even better pilot never saw combat. Belvin Womble Maynard was a native of the village of Morven, in Anson County. By late 1919, few would have denied him the right to be called "the greatest pilot on earth."[16]

His father, Dr. Atlas A. Maynard, closed his medical practice in Wallace when his health failed in 1905 and moved his family to Sampson County, where he hoped to regain his stamina. Settling in tiny Harrell's Store as a farmer and dairyman, he earned a decent livelihood for his wife and eight children. Belvin, born September 28, 1892, was the second oldest child. He attended Dell boarding school in nearby Delway, where he joined the baseball and debating teams, the band, and a singing quartet. He bought the first bicycle and automobile in the vicinity and, "a natural born mechanic," took responsibility for repairing all vehicles and machinery on the farm. It was said that at age seventeen he could take a car apart and put it back together.[17]

Dell School had Baptist support and functioned as a preparatory school for Wake Forest College. In addition to its primary departments, it operated a high school with well over one hundred students. There was a new brick building "with electricity and indoor plumbing." Children of Baptist ministers attended tuition-free. The catalog stressed Dell's objective to contend against "two of the most dangerous spiritual and social enemies of our time—selfishness and commercialism." It called Dell "one of God's means of making the world better and brighter, and mankind more useful." Every student was expected to devote two periods a week to "definite Bible study throughout the four year course."[18] Dell aimed high.

At the start of each fall term, the Maynards moved to a house across the road from the school and stayed until the academic year ended in May. Dr. Maynard, a Dell trustee, remained at Harrell's Store year-round to operate the farm but visited his family weekly. In the summer, the Maynards were all fully employed on their farm. A rare spectacle at Harrell's Store was the passage of a drove of mules headed for market. "We looked forward to the mule drives," said a family friend. "It was exciting seeing all the mules go by. . . . A big event." She described Mrs. Maynard as "extremely proper . . . a proud woman," who "sat up like a queen." After his fiercely evangelical education and upbringing, young Belvin Maynard, slim, handsome, and gifted, was ready to bear arms against a wicked world.[19]

He "secretly married" Essie Goodson, of Mount Olive, another Dell student, soon after they graduated in 1912. They taught for a year in South Carolina before he enrolled at Wake Forest College to study for the ministry. After two years there, he would be remembered as a student who worked at all kinds of jobs to pay his bills and support his wife and child. Maynard reportedly was ordained into the Baptist ministry and accepted a local pastoral charge. In June 1916 his bleak finances forced him to withdraw from the college and take a position as assistant foreman in a northern arms factory. When America went to war in 1917, he joined the army as a private; Essie and the children returned to Mount Olive, where she taught school for the duration of the conflict.[20]

Maynard was quickly transferred to the air service and trained before leaving for France. For eighteen months, he was a test pilot for every new plane sent to Issoudun's huge airfield, sometimes flying as many as twenty-two in a morning, and developed into a superb airman—too valuable in his present role to be released to the front. Maynard was "one of the few who can do a half barrel roll from the top of a loop." With unerring instinct, he could "fly at night and land without flares" or for two hours "without seeing the earth and come out right over his objective, perhaps 200 miles from where he started." His favorite planes were the Nieuport 28 and the big De-Havilland-4 two-seater, in which "he maneuvers and stunts as if it were a scout." He was also "a very skillful mechanic."[21]

Most American pilots were home by February 12, 1919, when Maynard, at France's Romerantin Airfield, set out on a Sunday to break the record of three hundred outside loops in sixty-six minutes. He had fallen into a habit of staging Sunday acrobatics, and a large crowd was on hand for his attempt. His Sopwith *Pup*, a fragile machine with a 160-horsepower Gnome Monosoupape engine, was said to be "the most difficult to handle" of all Allied planes. But its short fuselage made it sensitive to the controls and enabled the pilot to make tight loops quickly.[22]

A witness reported that at two thousand feet Maynard began to "cut a vertical circle against the blue background every twelve seconds. So consistently did he maintain his original altitude and time his loops that the performance took on a mechanical accuracy, his plane seemed to be following an invisible groove in the heavens." The strain on both man and machine seemed more menacing with each loop, but Maynard "churned on as if plowing endless furrows in the sky." "A thin trail of burnt-oil smoke described the plane's path," each loop overlapping the one before. After sixty-seven minutes and a remarkable 318 loops, he "circled the . . . field six times before his motor ceased . . . and began to cough for lack of fuel." After a neat landing, the pilot received an enthusiastic reception, including congratulations from General Mason M. Patrick, chief of the American air service in France.[23]

Still largely unheralded, Maynard returned to America in the summer, transferred to the aviation reserves, and prepared to reenroll at Wake Forest for the fall term. But almost simultaneously it was announced that aviation's first great postwar race, a round-trip between Mineola, New York, and Toronto, Canada, would be held in August. Savoring the chance to compete against other expert pilots, he chose to delay the return to his studies in order to take part in the race.[24]

Sponsored by the American Flying Club, this was a reliability race among various types of military planes — British, French, American, German, and others — piloted by both civilian and army fliers, fifty-two in all. It aimed to give "incentive to the commercial use of heavier-than-air machines among civilian aviators" and was touted as "the greatest race in aviation history." Those leaving Mineola's Roosevelt Field would fly a prescribed course over New York City and follow the Hudson River to Albany's Quentin Roosevelt Field. After a mandatory check, they would head, via Syracuse and Buffalo, for Toronto; others starting at Toronto would fly the reverse route. The round-trip of over a thousand miles was the longest race so far in aviation history.[25]

On August 25 twenty-eight planes left Mineola and the rest

Toronto, an array of French SE-5s and Spads, Italian Capronis, German Fokkers, Pfalzes and Albatroses, American-built DeHavilland-4s and Curtiss JN-4s, and a Hungarian plane. Among the pilots were ace Elliott Springs, three former members of the Lafayette Escadrille, and the highly regarded Carl Spaatz (commanding general of the U.S. Air Force by 1946). Civilians included Bert Acosta and other top fliers. The next day Sergeant C. B. Coombs, in a Curtiss, was the first from Toronto to reach Mineola and the first to leave on the return flight, though delayed by a five-hour rain. Storms at Syracuse and Buffalo forced down several fliers, while others became lost or otherwise delayed, creating confusion at both ends as to who was ahead and by how much. A name scarcely mentioned in early reports was that of Lieutenant Belvin Maynard.[26]

Working his way through frequent storms but having no trouble with his Libery-powered DH-4, Maynard managed an average of 134 miles an hour over the round-trip course. On August 30, he was named winner of the race, having made the 1,042-mile circuit in a record flying time of seven hours, forty-five minutes, and fifteen seconds. Suddenly, he was the toast of the air service.[27]

Lieutenant Belvin Womble Maynard of Sampson County. Courtesy of Caralee Maynard Rooks.

Though the "Flying Parson," as he was now known, was a household name, Maynard returned to his course work at Wake Forest on September 4; he intended to finish there, then go on to a theological seminary. His plan again unraveled—this time at news that the army would hold a Double Transcontinental Aerial Derby between New York and San Francisco in early October. As before, some planes would start at one end and some at the other end. There would be sixty-seven planes, including forty-four DH-4s, a DH-9A, two Martin bombers, seven SE-5s, a Spad, a Le Pere, an S.V.A., a Vought VE-7, a Thomas-Morse, and three Fokkers. Again, many of America's top pilots intended to compete, but now all eyes would be on Maynard.[28]

Westbound planes left Mineola on October 8 facing the unwelcome prospect of rain and gales the next day that would hamper visibility and turn fields to mush. Maynard's two-seater DH-4, the *Hello Frisco*, also carried his mechanic, William E. Klein, and a seven-month-old German police dog named "Trixie." The pup was a gift to Maynard from Winston-Salem's E. E. Wilson, a Dell School friend and fellow pilot, who brought her home from the war. She rode in the forward cockpit, a white scarf around her neck, "Snoopy"-style. Maynard had known Klein in France, where he was considered the best mechanic at the big Toul airfield.[29]

Departing Mineola at 9:25 A.M., Maynard got off to a fast start, making mandatory stops in Rochester and Buffalo before crossing Lake Erie to Cleveland. En route, he broke out of a wall of storm clouds east of Buffalo that caused four accidents during the race. The first man out of Cleveland, he built a lead crossing Lake Michigan that evening to Chicago. But he knew that Spaatz and others from the west were speeding east in good flying weather.[30]

After a restless night and no breakfast, Maynard left Chicago before sunrise on the second day, gaining three hours on his nearest westbound rival. But he soon met the roughest weather he had ever seen and was almost airsick from bumps so severe that his motor misfired. He briefly lost the lead but regained it by pressing on to Des Moines, Iowa, after sundown. The field there was so soft that, in landing, Klein had to leap onto the tail of the plane to keep it from nosing over. The next day there was more bad weather to Omaha, but the skies improved as he continued on in very cold air to North Platte, Nebraska, where he and Klein huddled over an oil heater to thaw out. While they were there, Spaatz and the two other eastbound leaders came in, having covered fewer miles than Maynard, so he knew that he was leading the pack. A blown tire cost fifteen minutes, but he reached Cheyenne at sunset in good weather. Seventeen planes had already crashed.[31]

At 5:30 A.M. on the fourth day, Maynard prepared to leave for Rawlins, Wyoming, hoping to reach San Francisco by nightfall. It was a forlorn hope. As the motor warmed up, there was a sudden loud noise and smoke from the motor blew back into the pilot's face. Thinking that it was an explosion and that the plane was on fire, he immediately stopped the engine. Klein climbed out and found that the radiator had burst. Maynard now showed his resourcefulness and expert knowledge of the Liberty motor. "It soon occurred to me," he wrote, "to try to blow through the overflow pipe running into the radiator. The . . . pipe was filled with ice and would not allow the water in the radiator to expand on becoming heated." But the radiator was too damaged for on-site repair. Borrowing a car, he raced to a Cheyenne garage and, despite a punctured tire, was back with the repaired radiator within an hour.[32]

The crisis averted, Maynard and Klein flew on to Rawlins, then to Salt Lake City and Salduro, Utah, the latter allowing a bath and a night's rest. Up again at sunrise on October 11, they stopped briefly at Battle Mountain, Reno, and Sacramento before reaching San Francisco, where the Presidio airfield was wrapped in fog. Circling for a glimpse of the field, Maynard at length found one and landed at 1:20 P.M. to a huge welcoming party of military personnel, photographers, and others. He had won the westbound race, but Spaatz and Emil Kiel reached Mineola only five and half hours later. The route was strewn with six dead fliers, all in wrecked DH-4s.[33]

Despite mandatory stops and other restraints, Maynard had crossed the continent in record time—three days, three hours, and five minutes. Ray L. Bowers writes that "a special aptitude for cross-country navigation," and "ability to handle perfectly the Liberty engine, stood behind Maynard's confident daring." Even Trixie won acclaim as the first transcontinental pooch.[34]

The nation was caught up in the drama of the race. Despite the carnage to airmen, a new age in aviation was clearly at hand. The *New York Times* emblazoned Maynard's arrival in San Francisco on page one, while the *News and Observer* and other North Carolina papers screamed reminders that he was a Tar Heel. But some questioned whether gains from such a race could possibly compensate for the loss in men and machines. By October 15 nine contestants were dead.[35]

Messages of praise for Maynard included one from "your parents and the people of Sampson County." He wired "God's country" a reminder that the East Coast was now only three days from the Pacific. "My . . . happiness . . . cannot be expressed. Again, thanks for your interest." There was also a telegram from Essie in New York begging him to find some other way to make the return trip. On the thirteenth race officials ordered him to leave San Francisco within forty-eight hours. Declining an invitation to dine with the visiting King Albert of Belgium, he took off on Tuesday, October 14, at 1:22 P.M., retracing his route and believing his lead to be almost insurmountable, though the Sierra Nevadas lay just ahead.[36]

The return flight began smoothly, Maynard's lead mounting as he passed over Nevada, Wyoming, and western Nebraska, averaging 169 miles an hour with the wind at his back. Leaving St. Paul, Nebraska, on Thursday, October 16, a plane, declared Maynard, "never carried two more jubilant pilots than old 'Hello Frisco' . . . that day." But in the afternoon, as they were sailing over eastern Nebraska at about two thousand feet, the motor began to miss. Maynard cut the ignition, looked below for a landing site, and glided to a perfect dead-stick landing in a cornfield. Klein got out, took off the distributor head, and had Maynard turn the propeller. The distributor was not working, but there was something much worse—no compression in the four cylinders. "He looked at me," wrote Maynard, "and pronounced . . . a death sentence: 'We have broken a cam shaft.' In reply I said: 'Well, I guess we are through.' "[37]

As the two men reviewed the situation, they reflected that they were near the town of Wahoo, about forty-five miles west of Omaha. The nearest camshaft was in Chicago, nearly five hundred miles east, and might as well have been in Addis Ababa. Wracking his brain, Maynard recalled hearing in San Francisco that Roy Francis's Liberty-powered Martin had crashed near Omaha.[38]

Rural Nebraska had never seen a whirl of activity equal to the next hours. Maynard ran to the nearest farmhouse with telephone wires and called Omaha to locate the wrecked Martin. He spoke with Francis himself, who said that his plane was at Yutan, eleven miles from the *Hello Frisco*, with a crew removing its motor as they spoke. Maynard called for an army truck to be sent from Omaha to Yutan. He rushed back to Wahoo and told Klein to start removing the heavy engine; the Nebraska Highway Commission agreed to send out a mechanic to help. Stopping the first car that came along, he persuaded the driver to take him to Yutan, where he found two mechanics removing the Martin's motor. A truck soon arrived from Omaha carrying soldiers to guard both planes. The Martin motor was placed on the truck and rushed over to Wahoo, where Klein was removing *Hello Frisco*'s engine.[39]

Excited Wahoo residents brought out a Delco portable lighting system that enabled the mechanics to work all night installing the Martin's motor. Others drove cars to the site, their headlamps providing additional light, while Maynard and Klein got some sleep in the front seat of a car. The motor transfer was a complex task that normally required "three full days with full hangar facilities." But before daylight, the replacement complete, Maynard ran the motor for a few minutes and found it working well. He took off at sunrise, having spent over twenty precious hours on the ground. Stopping briefly at Omaha, he rushed off to Rock Island in such a hurry that he left his maps behind and had to go back for them.[40]

At Rock Island, Maynard had lunch with evangelist Billy Sunday, who was conducting a revival there. As he approached Chicago, he ran into rough weather, the plane tossing "like a canoe in rough seas" and the motor misfiring. But he landed to the cheers of thousands, paused briefly, headed across Lake Michigan for Bryan, and reached Cleveland at sundown for a night's rest. Maynard and Klein left the next morning with the hope of reaching Mineola by 2:00 P.M. After short required stops at Buffalo, Rochester, and Binghamton, they set

**Artist's rendering of Maynard's force-down near Wahoo, Nebraska, where the pilot and his mechanic, William E. Klein, had a motor changed overnight during the Double Transcontinental Aerial Derby. Courtesy of Caralee Maynard Rooks.**

out on the last lap of the race, "happy as larks," Maynard wrote, "but not until my ship came within gliding distance of Roosevelt field did I feel perfectly satisfied."[41]

After circling briefly, the battered plane "oscillating like a 'falling leaf,'" Maynard landed at 1:50 P.M. EST and taxied toward a waiting crowd that included his wife and their four small children. Only nine other pilots finished the race; Spaatz had withdrawn at Buffalo on his flight west. Maynard's official flying time was computed at sixty-seven hours, three minutes, forty and one-half seconds, a new and dazzling record for a round-trip transcontinental flight.[42]

Maynard was America's new hero, "the premier flyer of the Army Air Service," as the *New York Times* put it. General Billy Mitchell recommended him for a "rating as a military aviator," with a 75 percent pay increase. Asked for comments, Klein declared that he would not work with anyone else as long as Maynard flew; he was "the greatest flyer in the service." The names of the ten pilots to complete the race were inscribed on the MacKay Trophy for 1919, "symbolizing high achievement in American aviation." The phrase "greatest pilot on

Maynard with his family after winning the Double Transcontinental race.
Klein is holding Maynard's younger daughter at right. National Air and Space
Museum, Smithsonian Institution, Washington, D.C. (SI Neg. 89-13765).

earth" was on many lips. A model of Maynard's plane was placed on
display at the Smithsonian Institution.[43]

To the charge of useless deaths, the army replied with tables
showing that, mile for mile, aerial racing was not as dangerous as
auto racing, though the latter took place "on specially prepared
tracks and with ample facilities for repairs and replacements." The
Double Transcontinental had a rate of under six deaths per 100,000
miles, whereas the mortality rate at the Indianapolis Speedway was
over 25 percent. On the average, the airplanes traveled forty miles an
hour faster than the cars.[44]

Within a day after the landing at Mineola, there was talk of Maynard making another flight across the country, this time in a "specially contributed mystery plane." This was a modified DH-4, its di-
hedral (V-shaped) wings straightened, extra gas tanks saddled to the
fuselage and fixed in the upper central wing panel, and a third wing

between the wheels. Carrying 306 gallons of gasoline, it was built to fly nonstop for 1,400 miles. Maynard was reported to have made test flights in it and mapped out a cross-country route.[45]

Meanwhile, newspapers commented that the contest boosted hopes for ocean-to-ocean mail. (In September 1920 the New York–Chicago airmail was extended to San Francisco.) Maynard noted that he had telegrams from over a hundred towns that hoped to build their own airfields. (The one under way near Winston-Salem was named in his honor.) General Mitchell saw the race as a signal that airpower, not seapower, would become America's first line of defense. No doubt remained, he said, that "complete control of the air by any nation means military control of the world." The transcontinental race pointed out the need for an efficient system of weather reporting and "an extensive communications apparatus." In addition, it showed design flaws in planes that must be corrected before they could serve the needs of either commerce or combat.[46]

Despite the Transcontinental's twelve deaths, the demand for books on aviation rose swiftly and a Philadelphian volunteered for the first flight to Mars. General Mitchell and other leading airmen were confident that Congress would agree to create an air force that was separate from the army. But support for such a change failed to materialize. The battleship would remain the keystone of national defense.[47]

Amid accolades over Maynard's achievement, the *Raleigh News and Observer* invited him to fly home and receive North Carolina's commendations. The town of Clinton, near his home, asked him to visit the county fair while he was in the state. Securing the army's permission, he left Mineola with Klein and Trixie in the *Hello Frisco* on October 31, refueled in Washington, D.C., and landed at Raleigh on the afternoon of November 1 on a field recently cleared by barnstormer Harry Runser two miles east of town. He had meant to stop at Wake Forest but, flying low over the school, found his gasoline too low to risk it, despite students "flocking out on the campus," he said, "in gym suits, and others, it looked like, in nature's garb."[48]

In Raleigh, Maynard was treated to a chamber of commerce luncheon, where future governor J. Melville Broughton called him "America's foremost pilot." Dr. William L. Poteat, president of Wake

Forest College, declared that Maynard had made the greatest contribution to flight, "with the possible exception of the Wright brothers." Maynard accepted an invitation to preach the next day at the
Baptist Tabernacle. Appeals for appearances came from Wilmington, Rocky Mount, Goldsboro, Rockingham, Winston-Salem, and
Fayetteville. The *Wake Forest Student*, a school publication, acclaimed
him "the first North Carolinian to become a world figure" and
planned an entire issue devoted to his exploits.[49]

On November 4 Maynard flew to Wake Forest with Governor
Thomas W. Bickett; it was the first flight for a North Carolina chief
executive. Finding no suitable field near the college, they returned to
Raleigh, and the next day Maynard left for Clinton. Reportedly, the
governor's flight persuaded many Tar Heel holdouts that planes
were, after all, a safe mode of travel.[50]

After Maynard performed some stunts for the county fair crowd,
the residents of Sampson County presented him a handsome silver
service; he then flew to his family home at Harrell's Store. On November 7 he went on to Winston-Salem for the dedication of Maynard Field, the city's new airfield seven miles out on the Kernersville
road. From there, he returned to Mineola.[51]

The army, seizing advantage of the publicity generated by the
transcontinental race, mounted an aggressive program to recruit
men for the aviation service, as so many had been discharged at the
end of the war. In North Carolina, planes from Langley Field (Hampton, Virginia) and the new Pope Field were criss-crossing overhead
to take part in Fourth of July celebrations, bond-raising campaigns,
fairs, and wherever else willing young men might be found. Maynard, of course, must have a hand in the campaign; in November the
army announced that he would fly a one-stop transcontinental
flight, beginning on November 26, apparently in the DH-4 "mystery
plane." But the plan was scrapped shortly afterward, and he was sent
on a low-profile mission to represent the air service at a convention
in Savannah, Georgia.[52]

In retrospect, this was a startling ill-omen. The apparent reason
for canceling the coast-to-coast flight stemmed from a late November talk Maynard had made to New York's Anti-Saloon League (ASL).
He was said to have stated that "'the failure of some of the pilots in

the recent transcontinental . . . race can be attributed to too much booze.'" Moreover, many of them went "to their deaths because they were flying drunk or with a hang-over." Drinking on duty, he declared, "is common among United States fliers. . . . I have seen men who always took a drink before flying."[53]

This was not all. In a November 30 sermon at Brooklyn Methodist Church, Maynard expressed shock over "the lack of clothes worn by women" at receptions in New York. "Raised in the peace, security, and modesty of country life," he disparaged "the costumes of women whose dresses are cut so low in the back that one can count the vertebra from the waist up." Moreover, women who frequented "cabarets and similar amusements are disinclined to assume the duties of motherhood." "There must," he said, "be a limit to this frivolity."[54]

Many urbane New Yorkers were offended by the gratuitous critique of their women and lifestyles from a Bible Belt airplane jockey. In response, New York City's health commissioner declared that low-necked dresses and high-heeled shoes were not immodest. The *New York Times*, also indignant at Maynard's presumption, argued

The "Flying Parson" poses at the pulpit. National Air and Space Museum, Smithsonian Institution, Washington, D.C. (SI Neg. 99-15354).

that he did not have the same liberty to speak out on matters of morality and women's dress as a person who filled a full-time pulpit. In an editorial, the paper commented that his "early training" may have made him see "declinations from absolute verticality as little different from complete prostration."[55]

Army bigwigs were outraged by Maynard's charge that drunken military airmen plied the skies between New York and California and that besotted fliers flew combat missions. In the campaign to attract new recruits, this was the worst publicity the service could receive. The army contended that all pilots in the Transcontinental had been observed at each checkpoint and no alcohol was reported. On December 11 General Charles S. Menoher, chief of the air service, summoned Maynard to Washington to explain himself. Major W. A. Robinson, commander at Sacramento's Mather Field, called the aviator's charges absurd. Commander George C. Westervelt, builder of the transatlantic seaplane NC-4, accused him of "indulging in ecclesiastical license. My experience," he said, "an extensive one, is that fliers leave liquor absolutely alone." There were rumors of a court-martial.[56]

Maynard declared that he had been misquoted, that his words had been taken out of context. He never intended to "criticize the air service, which . . . is made up of . . . young men who are noted for their sobriety." He had been dealt "the most serious injustice I have ever undergone." But the ASL asserted that his original typewritten statement "was subject to inspection by any one with a valid interest in the matter."[57]

Since he did not dare name names, Maynard had no defense against military and press attacks. It scarcely mattered that he may have met one or more drunken pilots in the course of the race or known service pilots who drank habitually. His charge, however worded, was a jab at the air service, one he could neither prove nor successfully retract. It was a hard way to confirm that Dell School's moral values were not synonymous with those of metropolitan America.

Maynard's army career, so brilliant at Thanksgiving, was essentially over before Christmas. On December 21 he told reporters that he would heed the advice of an eighty-year-old man who counseled him "not to make a fool of myself. . . . 'Go back to your pulpit and give up your flying and newspaper fame.'" He would leave the air service in January to "take charge of the little North Carolina church which he left to enter the service." He expected to have "little difficulty in obtaining his discharge."[58]

As for the alarming number of deaths and accidents in the transcontinental race, the army's view, set forth by General H. H. "Hap" Arnold, was that inexperience had led to pilot errors. He suggested that "more care should have been exercised in the selection of . . . pilots" for it.[59]

Maynard remained in the army until May 8, 1920. There was no more talk of the proposed transcontinental flight, but he was sent on a two-thousand-mile recruiting tour, evidently scheduled before the press debacle, through the South Atlantic states. He may have hoped that the tour would salvage his army reputation, but it did not. At the beginning of the tour he had a forced landing at Mineola, and on February 11, in the town of Aulander, in Roncommock's Bertie County, he broke a propeller trying to take off from a muddy field and had to wait six days for a replacement from Langley.[60]

In Fayetteville, on February 16, he took reporter Jane Myrover for a spin and then set out for Knoxville, Tennessee. Returning over the mountains, he became lost and landed at Morganton. Back in the air, he ran out of gas and was forced down at Hot Springs, in Madison County. He did not reach Asheville until March 30 after an unintended one-hundred-mile detour to Johnson City, Tennessee, then was delayed at Asheville for several days with engine trouble. He finally landed in Raleigh on April 7.[61]

Reporter Ben Dixon McNeill seized a chance to fly with Maynard in Raleigh but, though he had flown twice before, found it a harrowing experience. Maynard, angry, tired, and discouraged, swept with abandon south across the campus of the North Carolina College of Agriculture and Mechanic Arts. He "laid the tip of his lower right wing" on the laundry smokestack for what seemed to McNeill like "two or three eons." He climbed and dove at "tiny specks of people" with "horrified faces" and visible tonsils, pulling up at the last minute. The plane shuddered and, as Maynard descended, took out the top of a cedar tree. McNeill "aged many years" during the twenty-three minute flight and later accused the pilot of "heedless" recklessness.[62] The mystique was gone.

After his discharge, Maynard took a position with the Brooklyn Central YMCA, where he meant to stay a year or so before returning to Wake Forest. Moving his family to Queens, he also worked as an aerial photographer. On occasion, he introduced surprises, such as marrying a pair of passengers in flight in August 1922 and broadcasting an Easter sermon from his cockpit. He was about to join another flier to form an aerial photo company. Nothing had been heard in some time about his return to theological studies. If he had become cynical and embittered, the reason was obvious. What he had done *for* the army surely outweighed what he may have done *to* it.[63]

On September 6, 1922, Maynard performed with Trombley's Flying Circus at the Rutland, Vermont, fairgrounds, before a record crowd of twenty-five thousand. He took off in his Avro at 1:00 P.M. with two passengers. At two thousand feet, he did stunts, "working beautifully." The last was a stall-spin at eight hundred feet. But the plane "refused to respond and fell into a nose-dive," smashing into a field at the edge of the fairgrounds. He was the only one alive when taken from the wreck, but he died en route to the hospital. Some thought that he had misjudged his altitude when he began to dive. He had been scheduled to preach that evening at a local Baptist church.[64] There was no indication of suicide.

New York City honored the dead hero with a funeral procession through its streets. On September 7 Jonathan Daniels, son of the *News and Observer* editor, attended services "in the shade of a great grove" at the family's Harrell's Store home. There were no military honors, and no official representative of the state was present.[65]

"It was to Maynard's clean, Christian life," wrote Daniels, "rather than to his fame as an aviator," that the ministers testified in their eulogies. "They spoke of his young manhood, of his admirable life and character and referred with a certain lesser pride to his fame and ability as an aviator."[66]

Maynard's creed was embodied in a sermon he gave in Raleigh in 1919 before a congregation of 2,500. His text was from Matthew 5:16—"Let your light so shine before men that they may see your good works and glorify your Father, which is in heaven." Life is given us, he said, to glorify God and all that one does should be dedicated to that end. In the course of thirty minutes, he inveighed against broken homes, gossip, adultery, gambling, swearing, and, of course, Demon Rum.[67] The implied juxtaposition of his own celebrity and a censorious view of the weakness of his brethren went down well enough in Raleigh, but New York or Philadelphia might have been less charitable.

Ben Dixon McNeill, his view formed by a single unhappy encounter, once opined that Maynard "was not a gracious hero. He was pretty blunt with those who would touch the hem of his garment, very nearly as surly as Lindbergh." Like Lindbergh's, Maynard's aloofness may have been a necessary protection from an adoring public.[68] All but forgotten in his native state and slighted in air force annals, he had earned a better fate, despite his naive moralizing, than humiliation and imposed anonymity. Maynard may have been, as in the verse of W. H. Auden, one of those "runners whom renown outran / And the name died before the man," but he had a strong claim, in his time, to the title of "the greatest pilot on earth."

# The Lighthouse
## Keeper's Tale

WINGS FOR THE PEOPLE

In early 1919 Secretary of the Navy Josephus Daniels named Commander John H. Towers, the pilot on his first flight six years before, to plan an experimental transatlantic flight of navy planes. Towers mapped a route, recommended personnel, and assembled equipment. Other countries were pushing toward the honor of being first, but when the press found out about the project and questioned Daniels, he replied, "We hope to beat the world!" On a May morning, three Curtiss flying boats left Rockaway, Long Island, for the Azores. One came down and was rescued by navy ships waiting along the route; another landed at sea and taxied two hundred miles to the Azores. The third, the NC-4, reached the Azores nonstop on May 17 and Lisbon four days later; it was the first plane to cross the Atlantic—and it did so under Tar Heel direction.[1]

Whatever stature that Daniels may have gained, however, from his association with the flight of the NC-4 was lost in the Billy Mitchell affair. General Mitchell, a bellicose army spokesman for the view that airplanes had reached the point where they could sink battleships, staged a trial in 1921 in which planes did sink several ships. But many contended that the tests were inconclusive since the ships were unmanned, unarmed, and at anchor while under attack.[2]

In 1923 Mitchell won approval for another such experiment, this time off Cape Hatteras against the decommissioned battleships *New Jersey* and *Virginia*. The ships would be under radio-controlled sail and have their magazines filled with munitions; battle conditions would be simulated as nearly as feasible. Mitchell himself, who had flown to Hatteras several times over the past two years, now took a De Havilland there to test the sand and found that the beach could sustain the weight of a heavy plane. (On some of his visits he took up Hatterasmen who later boasted that they felt more comfortable in the sky than in a boat in the middle of Pamlico Sound.)[3] To fill up low spots along the shore for a smooth runway, Mitchell hired teams of local people and horses, including six-year-old Shanklin Austin and his mount at three dollars a day for himself and three for the horse. Islanders later recalled that Mitchell liked to sit on the counters of local stores and tell how he meant to sink the ships by dropping bombs down their smokestacks.[4]

Secretary Daniels was vehemently opposed to the test. If successful, it could lead to the formation of a separate air force and, in any case, would undermine the navy's prestige as America's first line of defense. The secretary declared that he would stand on the deck of any battleship Mitchell attempted to sink. "If he ever tries," said Daniels, "to aim bombs on the decks of naval vessels, . . . he will be blown to atoms long before he gets close enough."[5]

According to Mitchell's plan, thirty twin-engine Martin bombers would take off from the beach and fly out to Diamond Shoals Lightship, not far from the ships. They would attack the vessels at heights ranging from two thousand to ten thousand feet with bombs weighing up to a ton. But their bombsights were "simply a set of wires" in which the bombardiers "framed the target while allowing for wind, the speed of the aircraft—usually 75 to 80 miles an hour—and the type of bomb," a procedure infantry marksmen call "Kentucky windage." The Martins arrived at the beginning of September, "flying in a long string, like migrant geese, low over the beach." Along with the armada came barges loaded with fuel drums, tents, bombs, and supplies.[6]

On September 5, 1923, the planes flew out to a point near the lightship and began raining bombs on the *Virginia* and *New Jersey*.

The sturdy *New Jersey* held out for several hours before sinking, but the *Virginia* was dispatched in just twenty-seven minutes, a suggestive demonstration of what airplanes could accomplish. The great battlewagons took their places among the four hundred or more hulks lying in the "graveyard of the Atlantic." In the end, however, politics triumphed, and the air services remained firmly tied to the army and navy. Mitchell went down in history as an astute prophet of modern airpower, Daniels as a narrow partisan with obsolete views.[7] Daniels's staunch resistance to the experiments won him few military or civilian friends.

Progress in aeronautics under the pressure of war meant that the aviation industry, by about 1920, had passed beyond nearly anything that pastoral North Carolina could contribute to it. Needed now were technological sophistication, industrial muscle, and heavy capital outlay, none of which the state possessed. In the transition to peacetime, however, there remained a few veins of aeronautical ore that Tar Heels might yet mine. One was the quest for practical applications of the airplane, an area where creative thought might compensate for the state's material shortcomings. As sky swords were molded into hay forks, tinkerers might yet make some notable contributions.

With only a handful of airfields, North Carolina was ill-equipped to achieve even modest aviation goals. Surveys at the war's end indicated that the principal ones were the army's fields at Camps Pope and Greene (the latter a patch of cleared turf), Winston-Salem's Reynolds family and Maynard fields, Wilmington's Audubon, and the Pinehurst golf links. Pinehurst boasted a large white cross on the ground to guide pilots and a wind sock on a nearby dairy barn. "Emergency fields"—with stumps pulled and ditches filled in—included alfalfa fields near Fayetteville and Lexington; golf clubs at Durham, Wilmington, and Wilson; racetracks at Raleigh, Greensboro, and Salisbury; and a fine site southeast of Monroe. Others were at Rocky Mount, Hobgood, Hoffman, Newton, Clinton, Wadesboro, Dunn, and Council (in Bladen County). Fields at Gibsonville, Asheville, Greensboro, Coats, Smithfield, Raleigh, and a few other places rounded out the list. Few sites had so much as a hangar or a gas pump and were simply places to avoid crashing.[8]

In Wilmington in 1921, auto dealer Arthur Newkirk chased around Sunset Park in his "Aero-auto," a four-wheeler powered by an airplane motor and a rear propeller. It had no gears or brakes, starting and stopping by means of speed control. It was intended only for recreation, but the marriage of aeronautical and automotive technology, pursued in Wilmington since 1909, was a goal that would be sought by industry and society for the rest of the century—and more pressing with each decade.[9]

Twenty-seven-year-old Carr E. Booker of Raleigh had set out in 1919 to build the first "flivver" plane, which, like Ford's Model-T, would be so cheap to own and easy to fly that most families could afford one. Booker, holder of the state franchise for Crow-Elkhart automobiles, had a slight aviation background, but he drew up an impressive set of blueprints. He was also willing to spend all the money he had and could borrow to realize his ambition.[10]

His Hummingbird was a tiny biplane with an eighteen-horse Harley-Davidson motor. Sixteen feet long, with a wingspan of twenty feet, its 450 pounds made it the lightest plane ever built that "has the appearance of proving airworthy." The single-seater could cruise at sixty miles an hour and twenty miles per gallon. It would cost $12,000–$15,000 retail, the price of a medium-range car, and would have both commercial and private uses. A flight from, say, Raleigh to Little Washington, North Carolina, would cost only about $5, just a little more, and much faster, than driving. In 1919 Booker began work on a prototype at his Martin Street auto shop. It would need no airfield, its short takeoffs and landings requiring only a street or even a substantial rooftop. He enlisted the aid of barnstormers Harry Runser and Roscoe Turner to publicize it and of veteran army pilot Gus Leazar to test it.[11]

The Hummingbird was ready for public tests on April 13, 1921. Leazar had already flown it once but, at thirty feet, "it sagged and cracked," requiring repairs, enlarged wings, and other changes. Governor Cameron Morrison was induced to join the hundreds of spectators gathered to watch the next tests, and his young daughter cracked a champagne bottle on the propeller to christen the plane. Leazar took off at the "municipal airport" in the Longview section of town. The plane, says a witness, "climbed steadily . . . circled the

THE EYES OF THE INDUSTRIAL PIONEERS ARE TURNING
UPWARD TO THE SKY

The air must eventually become a great field of human endeavor and the popular-
izing of the air as a "playground" and servant of enterprise is the next great opportunity
for the far-sighted investor.

# Booker Aeroplane Corporation
## New York

AUTHORIZED CAPITAL—$200,000

All Common Stock, $10.00 per value per share.

Incorporated under the laws of Delaware.

The Booker Aeroplane

**Carr Booker's advertisement for his aviation company in Raleigh, 1920.
Courtesy of Karen Booker.**

field . . . and began to vomit black smoke. Then it came plummeting
. . . straight down. How it got itself levelled out and . . . landed none
can say." It had flown about 250 feet.[12]

Leazar tried again on April 23. Observers noted that the motor
sounded sluggish. He banked around, rose slightly, descended, rose
again, dropped, and recovered. He tried a turn but plopped onto a
pile of scrap iron in some bushes. He thought that he had given the
motor too much gasoline at the beginning. On April 26, before a

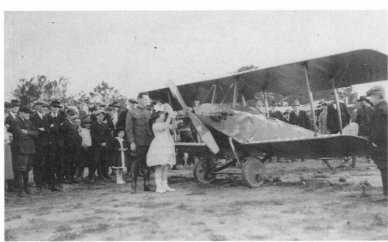

The small size of the Humming-bird is emphasized when it is viewed behind the left wing of a Jenny, Raleigh, 1921. Courtesy of Karen Booker.

Governor Cameron Morrison's daughter Angela christens Booker's tiny Hummingbird in Raleigh, 1921. Courtesy of Roy L. Booker.

handful of onlookers, he flew about a mile, returned, and landed on the field, where a broken wheel, dragging shock absorber, and cracked oil line caused a somersault. Unhurt, Leazar resolved to remain so by refusing to fly the plane again. Booker, nearly bankrupt, had spent a reported $55,000 on the project. He stored his plane for a time, then sold it to townsmen who owned a motorcycle garage. It never flew again.[13]

The death of the Hummingbird in Raleigh, 1921. Courtesy of Roy L. Booker.

A still more ambitious idea was that of John E. Parsons of tiny Mimosa in Polk County. In May 1922 he took to Asheville a model of an airplane that he had worked on for two years, embodying, he claimed, "many advantages over existing types." It was a passenger plane as long as a football field, with wings of "extraordinary length." He felt that he had succeeded "in applying to airplanes the principles exemplified in the wings of powerful flying birds" and hoped to find Asheville investors.[14] Just how he meant to power such a monster machine was not revealed. But he vanished as quickly as he materialized, without any public description, critique, or investment in his idea.

In New York in 1919, the Cantilever Aero Company, the brainchild of that dark genius, Dr. William W. Christmas, advertised "the first adaptation of the aeroplane to modern business." Photographs show a machine identical with his ill-fated Bullet but renamed the "Dark Horse." Its expected uses included overtaking ships as much as a day out at sea and dropping "a regulation mail-sack on her deck." This would prevent delays in port caused by late mail delivery. The Kerr Steamship Company had ordered one for this purpose.[15]

In April a Dark Horse was ordered by Mrs. S. E. J. Cox for the Texas Aero Club team that would compete in the September Gordon Bennett Trophy Race in France. Christmas was still promising 200-mile-an-hour speeds for his machine. Advertisements claimed that the company was building "flexible wing planes" for the race, a notice that continued to appear through August. But September brought news that Mrs. Cox would fly a different plane, Christmas having, apparently, failed to deliver. The ads had ceased altogether by 1921. It was the end of the Cantilever Aero Company but not, of course, Dr. Christmas.[16]

The inventor had lost none of his magic with admirers and backers. In 1922 the *New York Times* carried a letter from one of them, C. V. S. Wyckoff of New York City. Wyckoff's message was that there were two epochs in aviation history: the demonstration by Samuel P. Langley and the Wright brothers that flight was possible and the dis-

**Dr. William W. Christmas, 1929.**
From *Popular Mechanics*, August 1929, 234.

covery by Christmas of "the principle of bird flight and its application to airplanes." His flexible wing, as used in the Bullet, was "one of the greatest achievements in the history" of aviation.[17]

Though slowed by illness, Christmas was said by *Jane's All the World's Aircraft* in 1922 to be planning a "colossal" plane. His final version was a Flying Wing, a 72-ton monster 138 feet long and 31½ feet high, with a 262-foot wingspan, to carry 206 passengers and 17 crewmen—still rather small compared to that of John E. Parsons. It would have an eye-popping eight engines of 1,000 horsepower each, dining room, kitchen, lounge, and transparent leading edges in the wings through which passengers could view the horizon ahead. The seaplane version would add 30 feet to the wingspan. Just as astonishing, a group of Connecticut bankers would finance it. It would be easily the largest airplane in the world.[18]

The scheme died with the onset of the Great Depression, with no plane yet built, saving it from a more humiliating fate. Christmas died in New York City in 1960, at age ninety-six still inventing unbaked miracles like "Christmasite," an explosive second only to nuclear power. He was lauded by *Time* magazine and other periodicals as the first American after the Wrights to build and fly a plane and the source of about three hundred inventions, over one hundred of them in aviation. Patent Office records yield few of these and none that seem to have been commercially successful. Christmas died alone in hard-earned poverty. "It is too bad," remarks Arch Whitehouse, "that aviation history has given so little space . . . to Dr. Christmas's contribution to the science." In a flattering profile in *Popular Mechanics* in 1929, the author ventured that "the story of Dr. Christmas' life runs like a fairy tale"—a truer description than the writer could have known.[19]

## THE CLOUDLAND EXPRESS

Belvin Maynard's exploits created intense excitement in North Carolina but did not bring home to Everyman the sheer joy and acrobatic versatility of postwar airplanes. That remained the work of postwar barnstormers, notably Roscoe Turner and Harry Runser.

Turner had been an observation balloonist in the war, and Runser was a pilot who owned a surplus three-seater Canuck, a Canadian version of the Jenny. Neither man was a Tar Heel; they met in Virginia in the fall of 1919 and became partners, though Turner had never flown a plane. Their first exhibition was at the Caldwell County Fair in Lenoir in October, Turner serving as "grease monkey."[20]

Turner was soon doing wing stunts. W. F. Burgin of Elkin recalled riding a train to Winston-Salem in the fall of 1919 to see the show at Maynard Field. Turner and Runser had a sham dogfight with Frenchman Monte Rolfe, and Turner "shot" him down. Rolfe's partner jumped at five thousand feet, at first dropping like a rock. "Women screamed," according to Burgin, some fainting as Rolfe performed a variety of spins and flips, culminating in a "falling leaf dive" before landing smoothly, as did his partner.[21] It was already clear that the postwar airplane was infinitely superior to its predecessor of just half a decade earlier.

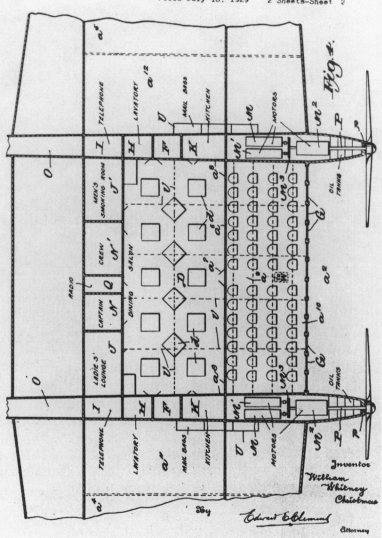

Patent for the Flying Wing, 1931. U.S. Patent Office, Washington, D.C.

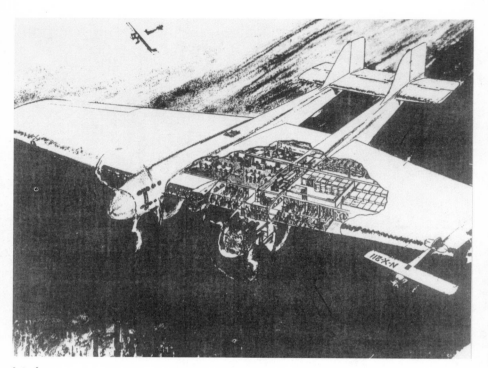

Burgin was at East Bend, in Yadkin County, a few weeks later, when Turner thrilled fairgoers "by walking the wings and swinging from the landing gear" as Runser circled the fairgrounds and town. On one such escapade, Turner's foot slipped off the wing as he climbed from his cockpit, "and [only] the presence of a guy wire and strut" saved him from defacing "the landscape and breaking up a perfectly good fair."[22]

Turner, who added new stunts at almost every show, was the delight of audiences. A tall, handsome, smiling man with a needlepoint mustache, he was the Beau Brummel of barnstormers, always dressed in tight riding breeches, blue military jacket with diamond-studded airman's wings, glossy boots, and gold and crimson goggles. Catnip to women, he spent most off-duty hours in their company. His professed career aim was to give the greasy business of barnstorming a touch of class.[23]

The Turner-Runser team soon reached Raleigh's state fair. Runser

Roscoe Turner (left) standing on his plane, the *Cloudland Express*, with Al Runser in the cockpit, Raleigh, ca. 1919. North Carolina Division of Archives and History, Raleigh.

pulled off his "falling in flames" act, involving a mile-long tailspin, simulating combat action, a surefire crowd pleaser. He was arrested for landing in W. S. Murchison's pasture just east of town but released after paying court costs. A bargain was struck with Murchison, but he soon plowed up half the field and the team shifted to a site four miles south of town. Turner and Runser appeared before the chamber of commerce to point out the need for a municipal field, and the chamber named a committee to pursue the matter. The airmen remained based in Raleigh for several months but gave shows as far afield as Rocky Mount, Henderson, Oxford, Roxboro, Durham, and other towns. At Asheville they parachuted and raced local flier Scott Dillingham. They "bombed" nearby Hendersonville with advertising leaflets and flew over the Blue Ridge at nine thousand feet. At each town they took up passengers at ten dollars a head, one joyful inebriate refusing to leave his seat until he had exhausted his fifty dollars. Turner was now spelling Runser at the controls.[24]

Looking for ways to increase their income, Runser briefly took a job as agent for Ace Trucks in Rocky Mount, making him North Carolina's first commercial pilot. Moving to Asheville in 1921, the airmen revealed a plan to establish a school for airplane and auto mechanics there. They canvassed prospects for manufacturing at Marion a plane called the Carolina Cloudster, probably a version of their Canuck. There was talk of a transcontinental flight to advertise the unrecognized glories of Greenville, South Carolina, and airmail-passenger-express service based in Charlotte. Their higher purpose was "to develop aviation in North Carolina and to show the public that airplane travel is just as safe as autoing."[25]

The two airmen parted company in 1921, Turner going on to become one of America's premier performers and racers. Having once been a circus lion tamer, he acquired a pet lion, "Gilmore," that flew with him. He returned periodically to North Carolina, always urging that towns build airfields and conducting himself so as "to have the spotlight of public interest turned upon him," writes his biographer, "and, through him, to aviation."[26]

Another well-known barnstormer of the time, also influential in boosting airfields and aviation across the state, was Alton Stewart of Coats, in Harnett County. Stewart was converted from auto mechanic to aviator when a Pope Field Jenny was forced down near his home in 1919. He persuaded pilots at Pope to give him flying lessons, earned a pilot's license in 1924, and was the first Tar Heel to receive a U.S. Department of Commerce license when they became mandatory in 1927. Meanwhile, he bought a Jenny and began taking up passengers for three-dollar rides, performing stunts at fairs, making charter flights, pulling political and advertising banners, giving lessons, and once flying newlyweds on their honeymoon. By 1926 Stewart was considered "one of the best commercial pilots in the South" and "a pioneer about whom something will be written 100 years from now." Like Turner, he was a tireless spokesman for aviation, published a newspaper insert entitled *Carolina Air News*, and behaved as a model of safe flying. He had moved with his family to Raleigh before a fatal crash in Harnett County on Christmas Day, 1929.[27]

**Opposite:
Barnstormer
Alton Stew-
art fueling
his Jenny,
1922.
Courtesy
of Hal
and Dave
Stewart.**

In 1924 Viola Gentry, of Rockingham County, became the first

female native of the state to learn to fly an airplane. She soloed in 1925 and, four years later, set the first nonrefueling endurance record for women with a New York flight of eight hours, six minutes, and thirty-seven seconds. A blue-collar aviator in a field of aristocrats, she never owned a plane but used most of her scant income as a cashier and clerk in stores and restaurants to rent them. Viola, surviving a catastrophic crash in pursuit of another endurance record, remained active for decades promoting aviation among youth groups, gathering materials for the History of Aircraft Research Center of the University of Texas, holding membership in the OX-5 Club and Silver Wings, and performing other roles.[28]

## BILL TATE'S CRUSADE

The only North Carolinian to stay in touch with the Wright brothers after 1911 was Bill Tate, their factotum from the earliest gliding days. Tate occasionally wrote to Orville Wright, always receiving friendly replies indicating his regard for Tate and his family.

Bill Tate harbored a deep concern that evolved into a personal crusade. After 1911 Tar Heels lost interest in the Wrights and felt no sense of pride in their state having hosted the first flight and helped to ensure its reality. To Tate, this was an injustice that had to be addressed. He began to explore ways to arouse interest in the first flight and those who made it possible. Aviation, he firmly believed, had its start in his Kitty Hawk front yard, where the first Wright glider was assembled.

Indicative of the problem he faced were the textbooks on state history. R. D. W. Connor's respected volume, *The History of the Old North State* (1906), might be excused for omitting the Wright brothers, but his *Makers of North Carolina History* (1911) also left them out, as did his 1916, 1923, and 1930 editions. D. H. Hill's *Young People's History of North Carolina* (1916) did not mention them either. Yet both writers were eminent academicians, Connor as the first secretary of the North Carolina Historical Commission and later the first archivist of the United States.[29] Not until Greensboro librarian Nellie Rowe's *Discovering North Carolina* (1933) was it disclosed in print

Opposite: Viola Gentry, a Rockingham County pilot who set the first endurance record for women. History of Aviation Collection, University of Texas at Dallas.

to schoolchildren that the Wrights made the first airplane flight at Kill Devil Hill. Even then, Hugh T. Lefler and Albert Ray Newsome accorded them only part of a paragraph in their authoritative state history of 1973, as did this writer in 1978 and 1983 junior high school textbooks.[30]

The indifference of the state's own chroniclers reflected the view that North Carolina contributed to aviation only sand on which to take off and land. The birth of flight was thus not an achievement this state could meaningfully claim. But there were sound arguments to the contrary. An example was the unrecognized role of scores of Tar Heels in the Wright brothers' success. Another was that North Carolina provided no more than a mat of leaves for the birth of Virginia Dare, the first white child born in America, though state publicists never tired of proclaiming her birthplace, despite its lack of historical significance.

Bill Tate was well qualified for his mission. He had known the Wrights longer and more intimately than anyone else in the state. He also had a better-than-average formal education at the Oxford Masonic Orphanage, Kitty Hawk High School, and Atlantic Christian Institute in Elizabeth City. An established civic leader in Dare and Currituck Counties, he was well known Down East. Photographs most often show him in his lighthouse keeper's uniform, his sober countenance and spare, erect figure suggesting a will of iron.[31]

Tate's crusade gestated for several years as he applied himself to earning a living. After 1902 he was employed at Martin's Point, north of Kitty Hawk, as manager of a sturgeon fishery, a sportsmen's club, and a sawmill, and he saw little of the Kill Devil Hill flying. He became a useful citizen as a member and, in 1907, chairman of the Currituck County School Board. He claimed credit for converting Currituck from Republican to Democrat but, though mentioned for the state legislature in 1910, did not run. In 1915 his party affiliations won him the post of keeper of Currituck's Long Point Lighthouse, where he remained until after President Woodrow Wilson's death.[32] As interest in the Wrights was rekindled in the 1920s, journalists found Tate the most articulate and well informed of the fliers' Outer Banks friends, and his name occasionally popped up in the newspapers. He came to be identified as North Carolina's unofficial spokesman for the Wrights, the person to turn to whenever the subject came up.

Wilbur Wright's death in May 1912 was Tate's call to action. Soon afterward, he wrote to Professor Connor at the North Carolina Historical Commission to urge that a monument be built at Tate's former Kitty Hawk home, but Connor seems to have shown no interest. Tate also broached the idea with Kitty Hawk friends but dropped it when he realized that some of them felt that he aimed only to enhance the value of his property. So France, in 1912, became the first country to build a memorial to the Wright brothers. A large black boulder, briefly inscribed, was erected in a field near Le Mans; eight years later the city displayed a five-foot marble shaft depicting Wilbur as a heroic nude with arms thrust skyward and named a street the Rue Wilbur Wright.[33]

Tate saw a new opening for his efforts in 1918, when he received an inquiry about the Wrights from Colonel Fred Olds. An avid Raleigh collector of historic documents and relics for the State Historical Commission, Olds was also a popularizer of state history through newspaper and magazine articles. After recounting briefly his experience with the Wrights, Tate observed in his reply that it was "surprising how little is known by us North Carolinians about the experiments and the development of the . . . aeroplane." Indeed, he ventured, one could ask "the first 100 people you meet" where and when the first flight was made, and at least ninety-nine would be unable to answer correctly. Even among those presumptively well versed in history, no more than 15 percent, by his estimate, could say. If, Tate declared, the first glider had been assembled in, say, California, "tourists would have come thousands of miles, just to see and stand on the spot." But his letter did not overtly seek to enroll Olds for the purpose of publicizing the Wrights' work, and all that came of it was a short article by Olds.[34]

A singular event that enhanced Tate's ties to the birth of aviation occurred in 1917. One day in January, a seaplane flew in low over Currituck Sound, its sputtering engine announcing a force-down. The pilot steered his crippled machine to a spot near Tate's lighthouse and ran it onto the beach. The airman was young Bennett D. "Ben" Severn, en route from New Jersey to Palm Beach, Florida. He hoped to complete the first flight of the plane, a Curtiss F-boat, down the East Coast between those two distant points. His summer business was taking Atlantic City sightseers on hops, and he wanted to use his off-season the same way in Florida. Trailing shortly behind, in a motorboat on the Inland Waterway, were four associates carrying fuel who would be available to help if the plane ran into trouble on its unexplored route.[35]

The airmen soon became acquainted with the Tates. Severn met Bill's daughter, nineteen-year-old Irene, a striking beauty, and they became mutually intrigued. He marveled to learn that she had known the Wrights before they became famous, that the first glider was assembled in her front yard, that she and her sister Lena had once worn clothes made from the first glider's wing cloth. Because Irene had attended high school and nursing school in Norfolk, Virginia, she had seen few planes since, perhaps, the Wright Model B in 1908. With the outbreak of war in Europe, aircraft rarely passed overhead, and a seaplane was a decided novelty.[36]

After his plane was repaired, Severn sent his companions on to Palm Beach while he stayed to pursue his courtship. The romance weathered an awkward moment when, as America entered World War I, overzealous Norfolk authorities arrested him on a charge of spying. Quickly released on Bill Tate's intervention, he married Irene at Coinjock Baptist Church on April 28, 1917. Many saw Irene's union with an airman as the fulfillment of a destiny,[37] whether Roncommock's or some other's.

Ben Severn's sojourn in Currituck had another lasting result. During their days on the sound, he and his crew were delighted to find skies and waters teeming with wildfowl. When the plane was fixed, they used it to hunt large numbers of ducks and geese. Local residents, and sportsmen who often traveled from distant points to hunt in the area, were irate over the destruction of birds in this way and protested to the state legislature. In March 1917 lawmakers enacted a statute banning the use of any "kind of flying machine in shooting, chasing . . . and harassing wild ducks, geese, swan or other wild fowl in and upon the sounds, rivers, creeks, or bays" of North Carolina. Conviction carried a fine of $100 to $500 or sixty days in jail. It was the state's first law dealing with aviation.[38]

Meanwhile, Severn's associates were hopping south from one coastal village to another, always the first fliers to visit these places. They lingered at Southport to take passengers on ten-mile flights at fifteen dollars each. On reaching Palm Beach, they had pioneered what would become one of the nation's most-traveled air routes.[39]

True to her destiny, Irene, after settling with her husband in New Jersey, took flying lessons and became a pilot. Over the years she logged over fifty thousand miles and was the first woman to fly round-trip between New York and Miami. It was with the Severns that Bill Tate enjoyed his own first flight in 1918.[40]

While visiting Currituck in March 1921, Ben Severn was again arrested, this time by the local sheriff, and charged with the theft of the $10,000 seaplane he was flying. Accused of taking it from Miami

Bennett Severn (left) and his crew display wildfowl they killed from their seaplane on Currituck Sound, 1917. This led to North Carolina's first law concerning aviation. Courtesy of Mrs. Willia Severn Raye.

without its owner's authority, he produced written permission to do so. The *Elizabeth City Advance* took his side, describing him as "a great big athletic fellow, young, strong, well-built and fine-looking," and "goodnatured and affable" into the bargain. He was soon released. The Severns's daughter Willia was named in honor of her grandfather, William J. Tate.[41]

## THE OBELISK AND THE PYLON

In 1922 Bill Tate renewed his long-interrupted correspondence with Orville Wright and mentioned Irene's marriage, a match for which the Wrights deserved some credit. He noted that she often visited the Currituck village of Coinjock on her flights and that his oldest son, Elijah, born in the winter of 1901, worked for the Curtiss Company as a motor mechanic. Wright replied fondly, recalling with pleasure the sight of the little Tate girls in their wing-cloth dresses.[42]

Tate decided that he might break down barriers to his crusade by publicizing himself. He ordered a business card that introduced him as "North Carolina's first contact with famous Wright bros. Original host and confidant of Wright bros. 1900–1903. Original North Carolina aviation booster."[43]

Over the years there had been occasional suggestions that America should erect a memorial to Wilbur and Orville Wright. Navy Secretary Josephus Daniels was among the first to raise the issue. In his newspaper in November 1913 he called for the creation of a Wright monument and museum, located, perhaps, in Washington, D.C. But Daniels and Tate had to wait thirteen years for the movement to take off. On December 17, 1926, First District congressman Lindsey Warren, acting on a suggestion by Elizabeth City editor W. O. Saunders (or, some say, Theo Meekins), introduced a bill for a Wright national park or monument to be erected at Kill Devil Hill. On the same day Hiram Bingham of Connecticut laid a similar bill before the Senate. The measure passed both houses in January 1927. Bill Tate exulted that his dream of a quarter century had been fulfilled.[44]

In a February speech to the House, Congressman Warren referred to Tate as keeper of the Wright flame and proud owner of first-flight relics. The speech transformed the Outer Banker into a national figure and influenced the rest of his life. Writing to Fred Olds in 1927, he noted that Warren had "read me into the congressional record so strong" that he was "flooded with letters from all over the U.S. and some from Canada." People sought his souvenirs and recollections, urged him to publish a book, and so on. He was trying to answer them, as more arrived almost daily, but was refusing requests for artifacts, having long since given away most of those he had owned.[45]

The U.S. Department of Commerce wanted him to compile data about Kill Devil Hill—for instance, subsoils and roads in the vicinity. The Wright legacy was becoming for Tate almost a full-time concern, though he continued to work at the Long Point Lighthouse. In May 1927 he informed Orville that he had recently visited the old campsite and found some building material, nondescript scraps that might now have historical value and appease supplicants. He also wrote again to Olds in Raleigh to ask his support for Tate's appointment as first keeper for the memorial lighthouse that Secretary of Commerce Herbert Hoover was planning for Kill Devil Hill. Tate thought himself "as well or better equipped with Local knowledge of the history of the coast" as anyone, and he had had a "close intimacy" with the Wrights. He hoped Olds would lobby North Carolina senators Lee Slater Overman and Furnifold M. Simmons and mention him in some of his writings. But Hoover's idea was rejected in favor of a pylon-shaped monument.[46]

Tate remained determined to have at least a small memorial in the front yard of his old home, by this time a Methodist parsonage. In association with Pastor W. A. Betts, he proposed that Kitty Hawk residents erect a monument on the site. This time he received an enthusiastic response. On September 26, 1927, he wrote Orville that the dedication would take place within sixty days, though the arrangements took longer than he expected.[47]

No outside funds were accepted for the monument, a small obelisk; a $75 check from a Californian was returned immediately. Donations by about three hundred local people—from a few cents to as much as $5—brought in $210. Pastor Betts wrote an inscription and publicized the project in newspapers. A block of Vermont marble was purchased, and an Elizabeth City firm carved the stone. Its modest inscription reads: "On this spot in 1900 Wilbur Wright assembled the first experimental glider with which the Wright brothers finally conquered the air. Erected by citizens of Kitty Hawk, 1927."[48]

Sadly, as Betts, in preparation for the ceremony, burned trash and leaves on the parsonage lawn on April 7, 1928, the house caught fire and was consumed by the blaze. Nevertheless, the obelisk was dedicated there on May 2 before a crowd of about five hundred people. Betts and Tate spoke briefly. Tate's grandson, Elmer R. Woodard Jr. (Lena's son), then unveiled America's first memorial to the Wrights. In December Bill Tate sent Orville a photograph of the marker and news that Kitty Hawk's high school had recently been renamed for the Wright brothers.[49] Obscure people who had once dined with the Wrights or helped cart some gear to Kill Devil Hill emerged as historical figures to be interviewed and widely quoted. Sundry "witnesses" began to come forward, often with spurious accounts of the first flight or tales of their intimacy with the Wrights.

**Bill Tate showing the Wright obelisk at Kitty Hawk.**
**Courtesy of Outer Banks History Center, Manteo.**

# Wright Memorial Tower to Go Atop Kill Devil Hill

The artist Verne C. Webster catches a vision of the Wright Memorial to be erected on top of the highest of the Kill Devil Hills as suggested by Wilbur and Orville Wright. The picture suggests the airport to be erected by the government as a part of the great memorial and gives us an unconventional pen portrait of Orville Wright himself. The inserts at the bottom of the picture represents the original Wright plane and Lindbergh's famous "Spirit of St. Louis." The sketch, done in crayon, was made exclusively

**Artist's conception of proposed Wright Memorial at Kill Devil Hill, 1928. The completed structure was rather different, and Charles Lindbergh's *Spirit of St. Louis* is unrecognizable. From *Elizabeth City Independent*, December 14, 1928.**

Events of the next two weeks completed the transformation of Tate's life. On May 20 Charles A. Lindbergh landed in Paris, generating a thunderous surge of global interest in aviation. The five first-flight witnesses, all still alive at the time, burst onto the national stage. As plans matured for the Kill Devil Hill monument, newshounds and others rushed to Kitty Hawk to learn more about the first powered flight from those who had seen it or, at least, claimed to have done so. Of its remote location, a shocked Orville Wright exclaimed, "Why no one will ever see it."[50]

As late as 1918, Kitty Hawk had no roads, only paths. But it now found itself in a spasm of real-estate deals and efforts to make the area accessible from the mainland for pilgrims to the Wright Memorial. Late 1927 saw plans for a Currituck Sound bridge to join Kitty Hawk to the mainland; a highway and bridge to link Kitty Hawk, Kill

Devil Hill, Nags Head, and Manteo; and progress on a Chowan River span at Edenton. Developers—on the eve of the Great Depression—bought up huge chunks of beach property for vacation lots. New Jerseyites Frank Stick and a partner, owners of the old Hayman Tract containing Kill Devil Hill, deeded part of it to the federal government for the monument and an adjacent airfield. The monument was dedicated in 1932.[51]

Tate kept Orville Wright abreast of the developments and mentioned his new Willys-Knight car; if Wright would come for the 1928 celebration of aviation's twentieth-fifth anniversary, the Tates would be pleased to entertain him. His family, he reminded the aviator, "put more faith in your experiments and the outcome of success than any other North Carolinians."[52]

The challenge that the new roads and bridges addressed was shown vividly in the December 1928 rites at Kill Devil Hill. In the *News and Observer's* view, there was "scarcely a more difficult place to get to in the state." Delegates and visitors via Norfolk faced a forty-mile drive south, with tractors and trucks standing by to free those vehicles that slipped off stretches of jerry-built planking. Overburdened ferries took passengers to Kitty Hawk, where five more miles of "doubtful road" led to Kill Devil Hill. Orville Wright was among the two hundred or so brave souls attending, his first visit since 1911. Bill Tate, "a natural orator," was a featured speaker. He also provided each official guest with a copy of a booklet of memoirs that he had compiled, at his own expense, about the Wrights' work at Kill Devil Hill.[53]

Over the years Tate wrote articles for newspapers and magazines and gave interviews to journalists and talks to audiences ranging from schoolchildren to aeronautical savants. As Orville's guest at an Engineers Club dinner in Dayton in 1933, he presented his host with the silver shovel with which Orville in 1928 had turned the first dirt for the Wright Memorial. In 1937 the "Capt. W. J. Tate Aero Club" was founded in New York to promote the education of boys seventeen or older who pledged to take flight tests for military service. A year later, he was the guest of Henry Ford at the opening in Dearborn, Michigan, of the transplanted Dayton bicycle shop and Wright homestead.[54]

Tate was also appointed to the Aviation Committee of the North Carolina Department of Conservation and Development, formed to "study the needs of the aviation industry" in the state and draw up plans for its "furtherance." The committee achieved little and Tate took scant interest in it, but his influence was seen in its concern that the *Wright Flyer* be returned from London and housed at Kitty Hawk. Orville had sent it abroad in 1925 because the Smithsonian Institution refused to recognize the Wrights' priority over Samuel Langley.[55]

In the 1942 edition of *Who's Who in Aviation*, the entry for Tate notes that for four years he had been a member of the board of governors of the National Aeronautic Association and its governor for North Carolina. He retired from government service in July 1940 and built a home in Coinjock, but his crusade was not yet concluded.[56]

In December 1940 he addressed the annual gathering at the Wright Memorial. He used the occasion to berate the Smithsonian for its mean-spirited and erroneous contention that the 1903 Langley plane was the first one capable of manned flight. Tate's tone of urgency in regard to the *Wright Flyer* reflected the concerns of the many people who feared that Nazi air raids would destroy the famous relic in the basement of London's Science Museum. His message on this occasion was uncharacteristically harsh. "It is a . . . humiliating condition," he said, "for Americans that this great exhibit . . . is lost to the nation forever. Think it over, and bow your head in shame that you have done so little about it." He wished "to put myself on record by saying we North Carolinians are lacking in civic pride. We are prone to pass unnoticed historical events that are monuments . . . and do nothing." But all was well when the plane, surviving the war unscratched, was returned to the United States in 1948, the Smithsonian at last conceding that the *Wright Flyer* was the first powered, heavier-than-air machine to carry a human aloft.[57]

Between 1938 and 1948 Orville and Tate exchanged letters about once a year, always at Tate's initiative. His terse commentaries mostly embraced the fortunes of himself and family and his efforts, as guardian of their legacy, to correct published errors about the Wrights. In 1943 he thanked Wright for a copy of Fred Kelly's new,

Captain Bill Tate in Coast Guard uniform. From William J. Tate, *Wings over Kill Devil* (1928), North Carolina Collection, University of North Carolina Library at Chapel Hill.

authorized biography, *The Wright Brothers*. His last two letters to Wright were written in October 1948 as the Dayton genius lay near death. "I always read your letters with delight," Wright replied, "Please write without waiting for letters from me."[58]

One of Tate's last public appearances was in January 1948 on *We the People*, a national radio show. Orville Wright died that month (within twenty-four hours of Kill Devil Hill surfman John T. Daniels), and Tate was an honorary pallbearer at his funeral. Addie Tate, who served Wilbur Wright his first Kitty Hawk meals and helped stitch the first wings, died in 1955.[59]

Bill Tate revered the Wrights as admirable and likable men who told good stories, ran about the beach behaving like birds, claimed to own a chicken that laid six eggs a day, and displayed other charming eccentricities. He was as vague about dates and incidents as he was about wingspans and coefficients—an exasperation to professional historians. In a 1927 letter to Orville, he recalled receiving from Samuel Langley in 1902 an inquiry as to "what you brothers were doing," whether "I had any pictures of the machine," and whether

Tate "felt like making a trip to Washington . . . that I would be doing the government a good turn. . . . Surely you can recall how we used to discuss it."[60]

Orville, recognizing that Tate's haziness could become an embarrassment, responded quickly to refresh his memory. Langley's telegram, asking permission to visit Kill Devil Hill, was directed, not to Tate, but to the Wrights. As they were about to leave for Dayton in 1902, Tate had handed them Langley's letter inviting them to Washington and asking them about "curved wings." His overture, though strictly ethical, was declined. Langley had not addressed Tate, and Tate's contention otherwise could create problems.[61]

In 1951 Fred Kelly, while compiling another book on the Wrights, sent Tate inquiries about the events of 1900 to 1903. Tate responded with tales like the Wrights' quest for fresh eggs from neighborhood housewives, his hosting the brothers at a goose dinner, his fear of their gasoline. The two men, born of sharply contrasting written and oral cultures, did not communicate well. Kelly complained to a friend that "Bill Tate always tells an apocryphal tale. At least half of what he told me was untrue." He added that it was the same or worse with the first-flight witnesses, who rarely told the same version of a story twice.[62]

Outer Bankers lived in a different world from Fred Kelly. They were unreliable when it came to figures, measurements, technical details, precise events. What remained was an afterglow of pleasant but dim memories of camp target practice, personal traits of the young Buckeyes, a spooked horse fleeing homeward, various accidents. Banker people were of little help in reconstructing what Kelly viewed as the truth about the airplane's invention. But his biographies contain little or nothing of the cheerful, productive relations between the Wrights and Bankers, the human interest and enjoyments of their daily interactions, the extent to which the Bankers made the first flight possible. He omitted most of what the surfmen would have thought worth cherishing about those interludes.

The myths and reminiscent trivia, however distorted by retellings and creeping old age, represent a valid part of the truth and deserve respect, recording, and a place in narratives of aviation's origin. Assembling and testing flying machines was a joyful and exotic experience; oral history enjoys a better press today than at midcentury. If technical prowess were all that mattered, no plane would have flown at Kill Devil Hill and a European would likely be credited with the first powered flight.

## THE WINGS OF MERCY

The men of the postwar Jennies were challenged to find useful work to pay the costs of maintenance and fuel. In October 1919 war veteran Edward L. Shealy brought private aviation to Charlotte when he landed his Jenny on a vacant lot in the Sedgefield section. He found a job with the *Charlotte Observer* flying newspapers to Gastonia, Kannapolis, Concord, and Davidson. Tarboro's W. H. Fillmore in January 1924 sprayed boll weevils as a government experiment, evidently without persuasive results. He also flew a doctor to Raleigh to give a radium treatment.[63]

War veteran Ralph Jordan flew at Morehead City as a fish spotter, and a few salesmen experimented, with indifferent results, in using airplanes to visit clients. Planes were not yet adapted to carrying cargo or produce, though small shipments could be made in open cockpits or containers. But such experiments failed to show that planes would become an important tool of commerce or public travel.[64] Barnstorming remained the best way to make money with small aircraft.

As the airplane slowly groped its way toward commercial respectability, it was fitting, if not fated, that the Outer Banks—the birthplace of flight—was the first region to benefit significantly from peacetime aviation. In the influenza epidemic that began late in the war and persisted for many months afterward, navy seaplanes flew from Hampton Roads to Outer Banks communities to transport patients to Norfolk hospitals. Aircrews also delivered "urgent prescriptions which could not be filled in Manteo" from Elizabeth City in twenty minutes, over a route that took twelve hours by water. Patients at Buxton, on Hatteras Island, who were eight hours by boat from medical assistance, were visited within thirty-five minutes by plane.[65]

Lieutenant Polk Denmark of Raleigh (standing) with Carteret County's Ralph Jordan (in cockpit) in France during World War I. Courtesy of James Denmark.

These aerial connections were indispensable to the well-being of Outer Bankers in the 1920s and later. No bridges yet connected the Banks to the mainland, and regular boat services reached only a few larger towns and seasonal resorts. Treatment for serious injury or illness, the delivery of babies, and rushing serum and assistance to isolated communities required seaplanes to prevent unnecessary deaths all along the Outer Banks.

In 1919 the U.S. Public Health Service designated Hampton Roads flying boats to send emergency physicians to coastal communities and soon extended the service to transport to Norfolk any patients needing emergency hospitalization. The service was taken over in

1922 by the Naval Air Station. From this point on, the navy furnished specially constructed and equipped F-5-L seaplanes and, later, X-P-B-2s, originally built for flights to Hawaii, and, at the time, the largest airplanes in the world. The flying ambulances brought back women in labor, accident victims, and those with life-threatening cases of pneumonia, influenza, appendicitis, heart attack, lockjaw, and other emergencies.[66]

There were many such flights, such as the one that carried a female patient from Hatteras to Norfolk in 1922, saving her life; many Banker children saw their first light in Norfolk hospitals. The airplane, said an observer, previously "associated with the idea of destruction of life, was converted to use of saving human life . . . in eastern North Carolina."[67]

The peril of these missions was dramatically shown by an incident in 1924. Two days before Christmas, the Norfolk Naval Air Station received from Buxton a wireless request to help a young man who was dangerously ill with pneumonia. Lieutenant W. B. Gwinn was the pilot, his F-5-L's open cockpits crammed with radioman, aviation mate, pharmacist, and crewman. They reached Pamlico Sound in about an hour, taxied to the beach, and added the patient, Andrew Williams, and his father, to the straining passenger list. The patient was tied securely in the forward cockpit beside pharmacist Gavins.[68]

As the plane gathered takeoff speed, its two Liberty engines groaning at their burden, a fishing stake gouged a large chunk out of the hull, though the plane still lifted off. Gwinn radioed the air station to have rescue vessels on hand when he reached Hampton Roads. While the plane swung down for its arrival, aviation mate William Mayo scrambled out onto the tail to help lift the plane's nose enough to prevent it from sinking. As the plane descended, Gavins threw away his chance of safety by trying frantically to free the patient from his restraints. Witnesses "saw the waters close over the plane, its propeller still revolving rapidly." Two crewmen fell into the propeller blades, one of them killed on impact; Gavins's skull was crushed and Andrew Williams drowned. It was the service's first fatal accident. But much benefit was rendered by this, the world's first air ambulance rescue service. In essence, it still exists.[69]

Jordan in France during World War I. Courtesy of James Denmark.

In 1930, after more than one hundred missions to the Outer Banks and now using a Boeing amphibian, the service remained the only one of its kind in the country "and perhaps the world."[70] Who would have guessed that, within fifteen years of the *Wright Flyer*'s takeoff, the airplane would find its first significant peacetime use in saving and benefiting the lives of the very people whose efforts made powered flight possible?

Fatefully or fortuitously, North Carolina was the site of aviation's first horizontal and vertical flights, for the achievement of which its sons and daughters labored mightily. Tar Heels built and experimented with airplanes as long as three decades before the Wright brothers. Before World War I, they designed hand-powered, steam, clockwork, electric, and gasoline aircraft. They also built gliders, monoplanes, biplanes, a multiplane, a helicopter, a dirigible, and, late in World War I, seaplanes.

An Oxford girl was the first woman to parachute from a plane and the impromptu inventor of the rip cord, a Roper man the first American to jump in a self-made parachute. Moreover, a Rockingham County resident was the first to set a woman's endurance flying record, a Greensboro man the first black to design a plane, a Charlotte African American the first of his race to patent an antiaircraft gun.

The state was the venue for fiction's first American balloon flight, the first novel inspired by successful airplane flight, and the birthplace of an early (possibly the first) poet of powered flight. Tar Heel artisans built a helicopter designed to drive and swim as well as fly; airplanes for roads and skies; a propeller-driven car; and planes meant to stand still in the air, carry up to a score of passengers, fly the first airmail, cost as little as cars, and fight powerfully in wartime. A Warrenton native was the first person to patent the aileron and cantilever construction, a Charlotte man one of the earliest national exhibition pilots.

Native sons flew in the border war with Mexico, the Polish-Russian War, in the pacification of the Riffs, and, in World War I, as instructors; balloonists; seaplane, pursuit, and bomber pilots; gunners; and observers. Tar Heel men were credited with the first plane ever shot down by an American airman and the first to make aerial attacks against Communist forces. The first transatlantic flight was commissioned by a Raleigh editor, and a Harrell's Store lad was arguably the greatest pilot of his age. Tar Heel craftsmen built the world's smallest plane and patented the world's largest.

With the marriage of Irene Tate of Kitty Hawk to a pilot, her aviation career, her brother's work for Curtiss, her father's crusade for the Wrights, and the launching, on the Outer Banks, of the first air ambulance service, early flight came full circle. The notes from Roncommock's reed, the muse that inspired William H. Rhodes's poetry, were borne on the same winds, over the same waters—once known as "the Sea of Roanoke"—that caress Salmon Creek on the west and Kill Devil Hill on the east. A great monument at Kill Devil Hill and a modest one at Kitty Hawk attest to the mighty transformation that began on these shores.

Those notes, that muse, circled the world many times before returning home to this storied land of the first English colony of 1585, and cradle, two years later, of the first English child born in America. Like aviation, America was founded, in spirit, on this same bleak shore. The visions of Walter Raleigh and Louis Mouillard, of Wilbur and Orville Wright and Luther Paul, of Tom Paine and Thomas Jefferson, had transposed themselves into stunning reality. Like America, the poet Rhodes's "landless, shoreless ocean of the sky" was conquered and humanity would reap the ever-expanding benefits of independence and democracy, as of powered flight. Roncommock's reed portended and ratified transcendent change: freedom for humanity both on the earth and beyond.

CHAPTER ONE

1. Lefler, *New Voyage to Carolina*.
2. Ibid., 39–40, 252–53.
3. Ibid., 226.
4. Mackworth-Praed, *Aviation*, 11.
5. Scott, *Shoulders of Giants*, 13; Mackworth-Praed, *Aviation*, 11–12.
6. Mackworth-Praed, *Aviation*, 3.
7. Schofel, *Hirum-Harum*.
8. Scott, *Shoulders of Giants*, 56; *San Francisco Golden Era*, January 27, 1867; *San Francisco News Letter*, March 9, 1867.
9. Powell, *North Carolina Fiction*, 128–29; Joseph C. Bates, *Bench and Bar of California*, 545–46.
10. Joseph C. Bates, *Bench and Bar in California*, 545–46; Powell, *North Carolina Fiction*, 128–29; Walser, *Literary North Carolina*, 18; "Sketch of the Late William H. Rhodes."
11. *San Francisco Golden Era*, January 27, 1867; Evans, "Steam-Powered Pioneer," 48–51; *San Francisco News Letter*, March 9, 1867, July 5, 24, 1869.
12. *San Francisco Golden Era*, January 27, 1867.
13. *San Francisco News Letter*, March 9, 1867.
14. Moscowitz, "Science Fiction of William Henry Rhodes," xiv–v; *San Francisco News Letter*, March 9, 1867.
15. *Elizabeth City Tar Heel*, October 24, 1902. Unless otherwise indicated, the newspapers cited herein were published in North Carolina.
16. Goldstein, *Flying Machine*, 176–79; Owen, "Frost at Kitty Hawk"; Frost, "Trip to Currituck."
17. Goldstein, *Flying Machine*, 179.
18. Jacobs, *UFO Controversy*, 5–10.
19. Ibid., 10–30, 68.
20. *Asheville Citizen*, April 15–16, 1897; *Goldsboro Headlight*, May 20 1897; *Norfolk Virginian*, April 8, 1897.
21. *Norfolk Gazette and Publick Ledger*, September 22, 1806.
22. *Wilmington Messenger*, April 6, 1897; *Wilmington Morning Herald*, April 6–7, 1897.
23. *Wilmington Messenger*, April 6–7, 1896.
24. *Wilmington Morning Star*, April 8, 1897; *Wilmington Messenger*, April 13, 1897.
25. *Wilson Advance*, April 15, 1897; *Fayetteville Observer*, April 10, 1897.
26. *Raleigh News and Observer*, April 16, 1897.
27. *Wilmington Messenger*, April 9, 30, 1897.

28. *Raleigh Daily Tribune*, April 17, 1897; *Lumberton Robesonian*, May 5, 1897; Hufham, "Who Carried Away the UFO?," 23–24.

29. *Goldsboro Headlight*, April 22, 1897.

30. *Wilmington Morning Star*, April 11, 1897.

31. *Wilmington Messenger*, April 7, 1897; *Chadbourn Journal*, April 8, 1897; *Sanford Central Express*, April 16, 1897.

32. Mackworth-Praed, *Aviation*, 113.

33. *Daily Charlotte Observer*, May 10, 1903. See also, e.g., November 25, 1901, January 27, 31, August 4, 14, 25, 1902.

34. *Daily Charlotte Observer*, May 10, 1903.

35. *Daily Charlotte Observer*, January 27, 1902, July 14, 1903.

36. *Daily Charlotte Observer*, March 1, 1901, May 17, 1903; *Norfolk Virginian-Pilot*, May 10, 1903.

37. *Norfolk Virginian-Pilot*, May 10, 1903.

38. Ibid.

39. *Daily Charlotte Observer*, May 17, 19, 1903; *Norfolk Virginian-Pilot*, May 19, 1903.

40. *Norfolk Virginian-Pilot*, December 18–19, 1903.

41. Verne, *Master of the World*; Verne, *Maitre du Monde*.

42. Walser, "Verne's Fantastic Voyages"; Verne, *Master of the World*, 108–14. For Rumbling Bald Mountain, see Powell, *North Carolina Gazetteer*, 439.

43. Verne, *Master of the World*, 11–29, 45–87.

44. Ibid., 244–45.

CHAPTER TWO

1. Gillispie, *Montgolfier Brothers*, 1–12.

2. Ibid., 13; Mackworth-Praed, *Aviation*.

3. Ibid., Becker, *Dreams and Realities*, 11–12; Mackworth-Praed, *Aviation*, 18.

4. Becker, *Dreams and Realities*, 13–16. A manned hot-air balloon allegedly ascended at Constantinople in 1702 but was not followed up.

5. Ibid., 34–40.

6. Ibid., 41–49; Mackworth-Praed, *Aviation*, 21.

7. Gillispie, *Montgolfier Brothers*, 50–64; Becker, *Dreams and Realities*, 17–20; Mackworth-Praed, *Aviation*, 22–23.

8. Mackworth-Praed, *Aviation*, 45–46.

9. Becker, *Dreams and Realities*, 52.

10. *Elizabeth City Tar Heel*, October 24, 1903.

11. *Norfolk Virginia Chronicle*, May 5, 1794; O'Dwyer Diary, October 12, 1825.

12. Murray, *Wake*, 395; R. H. Whitfield to "Uncle Nathan," November 22, 1848, Whitfield Papers. I am indebted to Albert Schneider for permission to cite this letter.

13. *Raleigh News and Observer*, July 28, 1883.
14. Keever, *Iredell*, 314–15.
15. *Raleigh News and Observer*, June 29, 1883.
16. Murray, *Wake*, 640.
17. *Weldon Roanoke News*, June 16, 1881.
18. *Wilson Advance*, April 22, 1897.
19. *Raleigh News and Observer*, October 16, 1890.
20. Ibid.
21. *New Bern Daily Journal*, February 4, 21–24, 1894.
22. *Daily Charlotte Observer*, April 6, 1882; *Concord Sun*, April 20, 1882.
23. *Daily Charlotte Observer*, August 8, 17–18, 24, 1889; *Kinston Free Press*, November 5, 1891; *Raleigh State Chronicle*, October 29–30, 1891.
24. *Raleigh Observer*, June 29, 1883.
25. Ibid.
26. Ibid.
27. *Raleigh Observer*, June 29, July 1, 1883.
28. *Raleigh Observer*, June 29, 1883.
29. *Raleigh Observer*, June 23, July 20, 1883; *Daily Charlotte Observer*, March 27, 1904; "Great Balloon Hoax."
30. *Daily Charlotte Observer*, March 27, 1904; *Raleigh Observer*, July 6, 1883.
31. *Charlotte Observer*, November 11, 1939.
32. *Charlotte Observer*, May 29, June 3, 1892.
33. *Charlotte Observer*, June 6, 22, July 16, September 16, 1892.
34. *Wilmington Messenger*, August 10, 13, 20, 1892; *Daily Charlotte Observer*, December 27, 1903.
35. Chatham Clark, *Wayward Balloon*, 11–14.
36. Ibid., 15–23.
37. Mackworth-Praed, *Aviation*, 26, 28–29, 58.
38. Ibid., 54–55.
39. Ibid., 96–97.
40. Shields, "Jacob Aaron Hill"; *Winston-Salem Journal*, March 9, 1902; *Reidsville Review*, May 20, 1902.
41. *Winston-Salem Journal*, October 20, 1968.
42. *Reidsville Review*, March 21, November 26, 1901.
43. *Reidsville Review*, February 18, 1902; *Daily Charlotte Observer*, November 25, 1901.
44. *Winston-Salem Journal*, March 9, 1902; Bishir et al., *Architects and Builders*, 206–8, 233–34.
45. *Winston-Salem Journal*, March 9, 1902.
46. *Raleigh Farmer and Mechanic*, May 6, 1902; *Reidsville Webster's Weekly*, May 8, 1902.

47. *Reidsville Webster's Weekly*, May 8, 1902; *Reidsville Review*, May 20, 1902.

48. *Winston-Salem Journal*, May 23, 1902.

49. *Winston-Salem Journal and Sentinel*, October 20, 1968; *Reidsville Webster's Weekly*, June 26, 1902.

50. *Reidsville Webster's Weekly*, June 26, 1902; *Reidsville Review*, June 26, 1902; *Winston-Salem Union Republican*, July 3, 1902.

51. *Reidsville Webster's Weekly*, March 5, 1903.

52. Photograph of Hill's dirigible, ca. 1902, in possession of his great-grandson, Reverend W. Stephen King, Pinnacle, N.C.

53. Ibid.

54. Reverend W. Stephen King, interview by author, June 25, 1992.

55. Ibid.; "Combined tobacco box and cutter, granted May 17, 1904," in *Annual Report of the Commissioner of Patents*, 1905, 24.

56. Reverend King, interview by author; Mrs. Margaret Kromm Federline, letter to author, June 21, 1992.

57. *Reidsville Review*, January 27, 1903; Ireland, *Entering the Auto Age*, 7–9.

58. *Reidsville Review*, February 18, 25, 1902.

59. Allott, *Jules Verne*, 201–6.

60. *Reidsville Webster's Weekly*, February 25, 1902.

61. Ibid.

62. Ibid.

63. Ibid.

64. *Reidsville Review*, July 18, 1902.

65. Ibid.; Donald Kromm, telephone interview by author, June 21, 1992.

66. Donald Kromm, telephone interview by author, June 21, 1992.

67. Corn, *Winged Gospel*, 7.

68. *Raleigh News and Observer*, October 14, 1908.

CHAPTER THREE

1. Mackworth-Praed, *Aviation*, 47; Scott, *Shoulders of Giants*, 3–20.

2. Gibbs-Smith, *Sir George Cayley's Aeronautics*.

3. Mackworth-Praed, *Aviation*, 60.

4. Ibid., 61; Becker, *Dreams and Realities*, 70–71.

5. Becker, *Dreams and Realities*, 78.

6. Mackworth-Praed, *Aviation*, 68–69.

7. Becker, *Dreams and Realities*, 144–45.

8. Ibid., 137.

9. Power, *Lee's Miserables*, 264–65; "Airship Inventor Offered to Carry the Mail."

10. Power, *Lee's Miserables*, 264–65.

11. Bishir et al., *Architects and Builders*, 193–239, 241–43.

12. *Sanford Central Express*, September 27, 1889.

13. *Wilmington Daily Journal*, November 15–17, 1871; *Salisbury Carolina Watchman*, June 9, September 12, 1881, April 20, 1882; *Daily Charlotte Observer*, June 3, 1886; *Charlotte Observer*, January 1, 1925.

14. Johnson and Stephenson, *Gatling Gun*, 121–28.

15. *Wilmington Morning Star*, October 10, 1871.

16. *Raleigh Daily News*, March 18, 1872.

17. Johnson and Stephenson, *Gatling Gun*, 119, 89, 103–4.

18. Parramore, "Background of Richard Jordan Gatling."

19. Johnson and Stephenson, *Gatling Gun*, 91–107, 150.

20. Ibid., 97–106, 119, 121, 124–25.

21. Ibid., 120.

22. Ibid., 120, 127–29.

23. Ibid., 129–34.

24. Ibid., 128.

25. Parker, *Ahoskie Era*, 176–77.

26. *Daily Charlotte Observer*, January 22, 1881.

27. Ibid; Unsigned, undated handwritten note in Daniel Asbury Biographical File; *Charlotte Observer*, March 31, 1918.

28. Bishir et al., *Architects and Builders*, 234; *Daily Charlotte Observer*, May 28, 1879; *Raleigh Observer*, July 11, 1884.

29. *Daily Charlotte Observer*, January 22, 1881.

30. *Salisbury Carolina Watchman*, April 7, 1881.

31. *Daily Charlotte Observer*, March 30, 1881.

32. Ibid.

33. *Raleigh Farmer and Mechanic*, April 21, July 14, 1881; *Philadelphia Evening Bulletin*, July 16, 1881, clipping in Daniel Asbury Biographical File.

34. *Daily Charlotte Observer*, March 18, 1882, February 12, 1904.

35. *Daily Charlotte Observer*, February 12, 1904.

36. Mrs. Melba Ambrose (cousin of John M. Smith), telephone interview by author, December 22, 1992; *Beaufort-Hyde News*, July 4, 1979.

37. *Beaufort-Hyde News*, August 30, 1979; *Manteo Coastland Times*, September 18, 1980, September 23, 1955; Rogers, *Home Grown*, 63; *Stonewall Pamlico Enterprise*, September 1, 1882.

38. *Manteo Coastland Times*, September 23, 1995.

39. *Beaufort-Hyde News*, August 30, 1979; Tate to Orville Wright, December 15, 1935, Wright Papers.

40. *Manteo Coastland Times*, September 23, 1955; Jordan Lilley, interview by author, Tyrrell County, N.C., February 20, 1994. William J. Tate, a Kitty Hawk friend of the Wrights, sent Orville Wright a clipping on Smith from the *Dare County Times*, n.d.

41. *Raleigh Farmer and Mechanic*, July 7, 1881; James Donelson Christmas (nephew of William Christmas), telephone interview by author, September 26, 1992; *Asheville Citizen*, May 16, 1889.

42. *Warrenton Gazette*, November 28, 1878.

43. Harold E. Morehouse, "Dr. William W. Christmas," manuscript dated December 28, 1970, Christmas Biographical File.

44. *Asheville Citizen*, May 16, 1889.

45. Ibid.; Luis Galvan to William Truettner, February 16, 1969, typescript, Christmas Biographical File.

46. Stearns, "Flying Wing," 234–35; Morehouse, "Dr. William W. Christmas."

47. Stearns, "Flying Wing," 234–35; Van Vliet, "Inside Story," 37, 60.

48. Whitehouse, *Early Birds*, 107; Christmas, "A Brief Authoritative Digest of the Most Important Works Done in Aviation by Dr. William W. Christmas," 1–7, Christmas Biographical File.

49. See, e.g., Wesley R. Smith to Thomas C. Parramore, January 28, 1995.

50. *Raleigh Farmer and Mechanic*, July 7, 1881; *New York Times*, January 10, 1885; Harmon, *Famous Case of Myra Clark Gaines*, 439–40; *Asheville Citizen*, May 14, 16, 1889.

51. *Washington (D.C.) Evening Critic*, June 27, 1881; *Washington (D.C.) National Republican*, June 27, 1881; *Washington (D.C.) Forney's Sunday Chronicle*, June 26, 1881; *Warrenton Gazette*, July 15, 1881; *Washington (D.C.) Post*, June 26, 1881.

52. *Washington (D.C.) Post*, June 26, 1881; *Oxford Torchlight*, July 19 1881.

53. *Asheville Citizen*, May 16, 1889.

CHAPTER FOUR

1. Tate, *Wings over Kill Devil*, 12.

2. Ibid., 3; *Raleigh News and Observer*, October 29, 1933.

3. Tate to Wilbur Wright, August 18, 1900, Wright Papers.

4. Ibid.; William J. Tate, "I Was Host . . . at Kitty Hawk," 191.

5. William J. Tate, "I Was Host . . . at Kitty Hawk," 191; Crouch, *Bishop's Boys*, 184–85; *Charlotte Observer*, December 17, 1928, August 26, 1934.

6. *Asheville Citizen-Times*, October 18, 1953; *Raleigh News and Observer*, December 9, 1943; *Elizabeth City Independent*, May 4, 1928.

7. Tate, *Wings over Kill Devil*, 14.

8. Ibid., 8; *Elizabeth City Independent*, December 2, 1937; William J. Tate, "I Was Host . . . at Kitty Hawk," 191; *Raleigh News and Observer*, January 12, February 16, 1948.

9. William J. Tate, "I Was Host . . . at Kitty Hawk," 192; Tate to Wilbur Wright, May 6, 1928, Wright Papers; Marden, "She Wore the World's First Wings," 27.

10. *Manteo Coastlander*, December 1978; *Raleigh News and Observer*, September 17, 1959; *Charlotte Observer*, October 28, 1928.

11. *Charlotte Observer*, October 28, 1928.

12. *Manteo Coastlander*, May 1973.

13. *Elizabeth City Independent*, April 27, 1927; *Charlotte Observer*, November 20, 1911. I am obligated for part of this description to Bill Harris, grandson of Elijah Baum.

14. *Manteo Coastlander*, May 1973.

15. Shanks, York, and Shanks, *U.S. Life-Saving Service*, 133–39.

16. *Elizabeth City Independent*, August 3, 16, 1928; *Manteo Coastlander*, May 1978.

17. *Elizabeth City Independent*, August 3, 1928; *Manteo Coastlander*, May 1973; Federal Population Census, Currituck County Population Schedule, Atlantic Township, 8. According to Bill Harris, Primitive Baptist Church and a second small school served the "up-the-road" population.

18. William J. Tate, "I Was Host . . . at Kitty Hawk," 189; *Manteo Coastlander*, May 1978.

19. Olds, "Mother of the Airplane."

20. Tate to Fred Kelly, October 28, 1951, Wright Papers; Crouch, *Bishop's Boys*, 190.

21. Haynes, *Wright Brothers*, 53; *Chapel Hill Weekly*, February 10, 1933 (contemporary's quotation); Albertson, *Wings over Kill Devil*, 16; Tate to Fred Kelly, December 5, 1951, Wright Papers (Tate's quotation).

22. Crouch, *Bishop's Boys*, 188–89.

23. Ibid., 196.

24. *Elizabeth City Tar Heel*, April 4, 1902.

25. Albertson, *Wings over Kill Devil*, 7–8.

26. *Elizabeth City Independent*, December 14, 1928.

27. *Charlotte Observer*, August 26, 1934, October 28, 1928, October 26, 1934; William J. Tate, "I Was Host . . . at Kitty Hawk," 189.

28. *Elizabeth City Advance*, October 28, 1928.

29. *Greensboro Daily News*, December 17, 1974; *Elizabeth City Independent*, November 18, 1932.

30. *Elizabeth City Independent*, November 18, 1932; Marden, "She Wore the World's First Wings," 27.

31. William J. Tate, "I Was Host . . . at Kitty Hawk," 189.

32. McAdoo and McAdoo, *Reflections on the Outer Banks*, 30–31; *Greensboro Daily News*, December 18, 1975.

33. Tate to Wilbur Wright, November 2, 1900, Wright Papers.

34. Ibid.; *Raleigh News and Observer*, February 12, 1948.

35. Tate to Wilbur Wright, March 8, 17, 1901, Wright Papers.

36. Ibid., March 17, June 9, 1901, July 20, 1902; *Charlotte Observer*, October 28, 1928; Hayman, "Hayman Clan," 5. According to L. D. Hayman, author of the family genealogy, the "old traditional" family graveyard was "right under

the sand dune on which the Wright Memorial . . . now stands" (pp. 13–14). Hayman adds that Captain Hayman was master of a sailing vessel when the Civil War began. Captured by a Federal ship, he volunteered as its "navigator" but promptly managed to run it ashore and escape. A strain of potato that he introduced from Barbados, considered superior to any local variety, is still grown on the Outer Banks. The captain was also a Methodist "exhorter," or Bible reader, and "tooth-puller" for his neighborhood when he met the Wrights.

37. Tate to Wilbur Wright, June 30, July 5, 1901, Wright Papers.

38. Ibid., March 17, June 9, 1901, Wright Papers.

39. *Charlotte Observer*, October 28, 1928.

40. *Raleigh News and Observer*, December 16, 1928.

41. Crouch, *Bishop's Boys*, 211–12.

42. Scott, *Shoulders of Giants*, 138–39.

43. *Charlotte Observer*, October 28, 1928.

44. *Greensboro Daily News*, December 17, 1974; unidentified clipping entitled "Albemarle Sound Region and the Outer Banks of North Carolina, 1738–1939," John C. Emmerson Jr. Scrapbook, p. 624 (hereafter cited as Emmerson Scrapbook); *Elizabeth City Independent*, November 18, 1932.

45. Unidentified clipping, Emmerson Scrapbook; McAdoo and McAdoo, *Reflections on the Outer Banks*, 31; *Raleigh News and Observer*, August 1, 1926.

46. *Raleigh News and Observer*, August 1, 1926. See also unidentified clipping, Emmerson Scrapbook, and McAdoo and McAdoo, *Reflections on the Outer Banks*, 31.

47. Unidentified clipping, Emmerson Scrapbook.

48. "Kitty Hawk Twenty-five Years after the Event," 30; unidentified clipping, Emmerson Scrapbook.

49. Tate to Wright, October 12, 1901, Wright Papers.

50. Ibid., January 15, 1902.

51. Ibid., January 24, 1902; *Manteo Coastlander*, May 1973; Crouch, *Bishop's Boys*, 235.

52. *Greensboro Daily News*, August 4, 1971.

53. *Charlotte Observer*, October 28, 1928.

54. *Elizabeth City Advance*, December 16, 1937.

55. Drinkwater, "I Knew Those Wright Brothers Were Crazy," 289; Crouch, *Bishop's Boys*, 235–36.

56. Crouch, *Bishop's Boys*, 238–39.

57. *Elizabeth City Independent*, November 18, 1932.

58. *Elizabeth City Independent*, May 4, 1928; Tate to Fred Kelly, October 28, 1951, Wright Papers; *Norfolk Virginian-Pilot*, December 15, 1950.

59. *Elizabeth City Tar Heel*, October 3, 1902; *Raleigh News and Observer*, December 16, 1928.

60. *Elizabeth City Tar Heel*, October 3, 1902.

61. Ibid.

62. *Elizabeth City Tar Heel*, July 11, 1902.

63. McFarland, *Papers of Wilbur and Orville Wright*.

64. Ibid.; Albertson, *Wings over Kill Devil*, 4.

65. Tate to Fred Kelly, October 28, 1951, Wright Papers; unidentified clipping, Emmerson Scrapbook.

66. Albertson, *Wings over Kill Devil*, 9.

67. *Raleigh News and Observer*, December 16, 1928; Arthur, "World's First Airplane Casualty," 11.

68. Carrie Rossi, "Flyer Three," 3–4, typescript, J. T. Daniels Biographical File.

69. McFarland, *Papers of Wilbur and Orville Wright*, 2:281–84; *Elizabeth City Advance*, December 17, 1978.

70. Tate to Wright, April 29, 1903, Wright Papers.

71. Ibid., June 30, 1903.

72. *Raleigh News and Observer*, December 16, 1928.

73. *Raleigh News and Observer*, January 16, 1938; *Charlotte Observer*, October 18, 1928; McFarland, *Papers of Wilbur and Orville Wright*, 1:358–78.

74. *Raleigh Morning Post*, November 13, 1903; McFarland, *Papers of Wilbur and Orville Wright*, 1:245.

75. Drinkwater, "I Knew Those Wright Brothers Were Crazy," 289.

76. Ibid.; *Norfolk Virginian-Pilot*, December 20, 1903.

77. Drinkwater, "I Knew Those Wright Brothers Were Crazy," 289; *Charlotte Observer*, October 28, 1928.

78. *Norfolk Virginian-Pilot*, December 9, 1903.

79. *Norfolk Virginian-Pilot*, December 20, 1903; Walsh, *One Day at Kitty Hawk*, 139.

80. *Charlotte Observer*, December 24, 1934.

81. *Eden Daily News*, December 13, 1970.

82. Unidentified clipping, December 26, 1943, J. T. Daniels Biographical File; *Raleigh News and Observer*, December 16, 1928.

83. *Raleigh News and Observer*, April 23, 1961.

84. *Washington (D.C.) Evening Star*, December 18, 1951; Aycock Brown, *Birth of Aviation*, 15; *Greensboro Daily News*, December 17, 1927; *Charlotte Observer*, December 16, 1928.

85. Arthur, "World's First Airplane Casualty," 12.

86. *Winston-Salem Journal*, December 17, 1935; *Elizabeth City Independent*, November 18, 1932; *Wilmington Daily Journal*, February 9, 1927.

87. *Raleigh News and Observer*, December 27, 2000; *Greensboro Daily News*,

December 17, 1974; *Elizabeth City Advance*, December 16, 1937; *Eden Daily News*, December 13, 1970.

88. *Elizabeth City Advance*, December 16, 1937.

89. *Norfolk Virginian-Pilot*, December 15, 1950; *Charlotte Observer*, October 28, 1928.

90. *Elizabeth City Daily Advance*, December 16, 1927; *Eden Daily News*, December 13, 1970.

91. *Manteo Coastland Times*, December 11, 1953.

92. *Raleigh News and Observer*, December 16, 1928.

93. *Elizabeth City Independent*, November 18, 1932.

94. *Raleigh News and Observer*, December 18, 1948; *Elizabeth City Independent*, February 1, 1929; *Boston Christian Science Monitor*, December 18, 1943; Sharpe, "Johnny May Go to Washington." On the morning of February 29, 1952, Moore, the last living witness to the first flight but in poor health, blew his head off with a twelve-gauge shotgun. *Raleigh News and Observer*, February 29, 1952.

95. *Raleigh News and Observer*, November 18, 1932.

96. Crouch, *Bishop's Boys*, 286.

97. *Elizabeth City Independent*, November 18, 1932; *Charlotte Observer*, October 28, 1928.

98. *Raleigh News and Observer*, December 16, 1943.

99. Jakab and Young, *Writings of Wilbur and Orville Wright*, 55.

100. *Elizabeth City Independent*, December 16, 1937; *Charlotte Observer*, February 2, 1937.

101. *Elizabeth City Independent*, November 18, 1932.

102. *Charlotte Observer*, October 28, 1928.

CHAPTER FIVE

1. *Norfolk Virginian-Pilot*, December 23, 1903.

2. Chambers, Shank, and Sugg, *Salt Water and Printer's Ink*, 242–43; Moore, "Reporter Who 'Scooped' the World," 1281–85; *Charlotte Observer*, December 18, 1938; *Norfolk Virginian-Pilot*, December 11, 1949; C. C. Grant affidavit, April 11, 1929, in "Clippings, 1903–1910," Wright Papers. An Associated Press report of 1953 says that Moore hired Daniels, Etheridge, and J. J. Dosher to let him know if the Wrights flew. Shank's Raw Materials, vol. 4, 1282, Shank Papers.

3. *Charlotte Observer*, December 18, 1928.

4. *Charlotte Observer*, December 25, 1903; *Norfolk Virginian-Pilot*, December 18, 1903.

5. Tate to Wilbur Wright, December 26, 1903, Wright Papers.

6. Ibid.

7. Crouch, *Bishop's Boys*, 307, 314–15, 278–86, 301–15.

8. Tate to Wilbur Wright, August 4, 1906, and Adam Etheridge to Wilbur Wright, November 15, 1906, Wright Papers.

9. Crouch, *Bishop's Boys*, 344–45.

10. Ibid., 324.

11. Tate to Wright, April 1, 1908, Wright Papers.

12. Crouch, *Bishop's Boys*, 354.

13. *Elizabeth City Advance*, February 27, 1935.

14. Kramer, "Kramers," 45, Kramer Family Papers.

15. Ibid., 47–48.

16. Crouch, *Bishop's Boys*, 354–55.

17. Wilbur Wright to George Spratt, July 14, 1908, copy in McSurely Papers; Crouch, *Bishop's Boys*, 355–56.

18. *Charlotte Observer*, April 30, 1908.

19. Crouch, *Bishop's Boys*, 355–57; *Charlotte Observer*, June 12, 1928.

20. Crouch, *Bishop's Boys*, 358–59; *Raleigh Spectator*, August 18, 1985; Drinkwater, "I Knew Those Wright Brothers Were Crazy," 294.

21. *Elizabeth City Tar Heel*, May 29, 1908; *Charlotte Observer*, May 16, 1908; Crouch, *Bishop's Boys*, 358–59.

22. *Elizabeth City Tar Heel*, July 29, 1908.

23. *Greensboro Patriot*, May 13, 20, 1911; *Raleigh Caucasian*, May 14, 1908; *Concord Times*, May 19, 1908.

24. *Charlotte Observer*, April 30, May 5, 1908.

25. Crouch, *Bishop's Boys*, 366–70.

26. Ibid., 373–76.

27. Etheridge to Wilbur Wright, August 31, 1909, May 9, November 14, 1910, March 10, 1911, and Wilbur Wright to Etheridge, September 23, 1909, November 29, 1910, Wright Papers.

28. *Elizabeth City Advance*, September 21, 1911.

29. Crouch, *Bishop's Boys*, 426–33.

30. *Elizabeth City Advance*, October 13, 1911; Crouch, *Bishop's Boys*, 443–44.

31. *Raleigh News and Observer*, October 18, 20–21, 1911; *Henderson French Broad Hustler*, October 10, 1911.

32. *Raleigh News and Observer*, October 7, 11, 14, 17, 1911.

33. *Charlotte Observer*, October 12, 1911.

34. *Charlotte News*, October 20, 1911.

35. *Charlotte News*, October 20, 25, 1911; *Charlotte Observer*, October 24, 1911.

36. *Greensboro Daily News*, October 25, 1911.

37. *Raleigh News and Observer*, October 25, 1911; *Spencer Crescent*, November 24, 1911.

38. *Greensboro Daily News*, October 27, 1911.

39. *Wilmington Morning Star*, October 27, 29, 1911; W. J. Tate to Wilbur Wright, April 1, 1908, Wright Papers; *Charlotte Observer*, October 12, 1911; *Elizabeth City Advance*, October 27, 1911. Besides the lifesavers, the work crew included Orville, Lorin, and Horace Wright, and Alexander Ogilvie.

40. Wilbur Wright to Adam Etheridge, August 15, 1911, and Adam Etheridge to Orville Wright, August 16, 1911, Wright Papers.

41. Adam Etheridge to Orville Wright, August 16, 1911, Wright Papers; *Raleigh News and Observer*, December 17, 1939.

42. *Norfolk Virginian-Pilot*, June 12, 1927.

43. *Raleigh News and Observer*, October 7, 1911.

44. Crouch, *Bishop's Boys*, 447.

45. *New Bern Daily Journal*, November 28, 1911; *Elizabeth City Independent*, December 2, 1927; *Charlotte Observer*, October 16, 1911; M. D. Hayman to Allen R. Hueth and Frank Stick, October 5, 1927, Dare County Record of Deeds, 1926–27, 5:321. This apparently is the Kill Devil Hill property that Stick, after Hueth's death, deeded to the federal government for the Wright monument.

46. *Greensboro Daily News*, October 13, 1911.

47. *New Bern Journal*, November 28, 1911.

48. Ibid.

49. Ibid.

50. Ibid.

51. Ibid.

52. Ibid.

53. Etheridge to Lorin Wright, November 12, 1912, Wright Papers.

54. *Elizabeth City Advance*, October 13, 1912.

CHAPTER SIX

1. Liberatore, *Helicopters before Helicopters*, 6, 9.

2. Ibid., 12.

3. Ibid., 22–23.

4. Ibid., 32.

5. Ibid., 2.

6. Ibid., 30, 50; Becker, *Dreams and Realities*, 11.

7. Liberatore, *Helicopters before Helicopters*, 73–76.

8. Mackworth-Praed, *Aviation*, 135; Editor's insert note, Parramore, "Flight of the 'Bumble Bee'," 8.

9. Parramore, "Flight of the 'Bumble Bee'," 8.

10. Grayden Paul, "Grayden Paul Reminisces," 1:66–67.

11. Aycock Brown, "All in One Lifetime," copy furnished to me by Professor Charles L. Paul; Grayden Paul, "Grayden Paul Reminisces," 66–67.

12. Grayden Paul, "Grayden Paul Reminisces," 67.

13. Dudley, *Carteret Waterfowl Heritage*, 97–98. This work was first published serially in *Decoy* magazine, Burtonsville, Md., in 1993.

14. Paul to Edison, March 10, 1913, copy, and Edison to Paul, March 1913, letter, in possession of Grayden Paul Jr.

15. Bensen with Abbott, *Dream of Flight*, 17.

16. Frances Rogers, "Luther Paul," copy furnished to me by Professor Charles L. Paul.

17. Selby and Spencer, "Fairfield," 37–38.

18. Paul to National Inventors Council, February 14, 1942, typescript copy in possession of Grayden M. Paul Jr.

19. Verne, *Master of the World*; Liberatore, *Helicopters before Helicopters*, 73–87.

20. Paul to National Inventors Council, February 14, 1942.

21. Ibid., March 23, 1942.

22. Ibid.

23. Ibid.

24. Ibid; Rogers, "Luther Paul"; Lina Bilar, "William Luther Paul," 5. I am indebted to Ms. Bilar, a granddaughter of Luther Paul, for a copy of this undated sketch, which she presented in a talk to the Beaufort Historical Society.

25. Bilar, "William Luther Paul," 5; Grayden Paul, "Grayden Paul Reminisces," 70.

26. Ibid.

27. Ibid.

28. Ibid; Daphne Paul, "Accomplishments of William Luther Paul," 87–88.

29. Barden to Aycock Brown, February 10, 1942, letter, in possession of Grayden M. Paul Jr., to whom I am indebted for a photocopy of the original; Paul to National Inventors Council, copy, March 23, 1942.

30. Green to Paul, July 14, 1942, letter, in possession of Grayden M. Paul Jr.; Liberatore, *Helicopters before Helicopters*, 75–76.

31. Liberatore, *Helicopters before Helicopters*, 75.

32. Paul to National Inventors Council, March 23, 1942; Mrs. Louise Bedworth, interview by Professor Charles L. Paul, December 28, 1997, January 3, 1998; Grayden Paul, "Grayden Paul Reminisces," 70. I am indebted to Professor Paul for a written summary of his interview.

33. *Raleigh News and Observer*, December 18, 1937.

34. Paul, "Grayden Paul Reminiscences," 67–70; Liberatore, *Helicopters before Helicopters*, 115–20.

35. Unidentified newspaper clipping, August 21, 1909, Aviation File, Cape Fear Museum. See also *Wilmington Morning Star*, December 17, 1911.

36. *Durham Morning Herald*, September 28, 1911; *Wilmington Morning Star*, December 17, 1911.

37. Robert T. Herbst, telephone interview by author, June 3, 1996.

38. Booker, "Wright Brothers Plane in a British Museum: Why?," 3–4.

39. Liberatore, *Helicopters before Helicopters*, 49, 97.

40. *Raleigh News and Observer*, February 1, 1920.

41. Bensen with Abbott, *Dream of Flight*, 4–7, 17–19.

42. Ibid.

CHAPTER SEVEN

1. Mackworth-Praed, *Aviation*, 133, 143, 150.

2. Corn, *Winged Gospel*.

3. Ireland, *Entering the Auto Age*, 1–8; Watson, "Sailing under Steam," 33.

4. *Official Gazette of the United States Patent Office*, 1910, 751.

5. *Charlotte Observer*, June 16, 1910; *Wilmington Morning Star*, November 16, 1910; *Raleigh News and Observer*, June 6, 1910.

6. *Charlotte Observer*, June 16, 1910; *Greensboro Daily News*, November 15, 1910.

7. *Charlotte Observer*, June 16, 1910; Hall, "Another First Flight," 8. The trial flights may have been canceled because of an accident or because tests showed that the machine was not satisfactory—a happy outcome, given the humiliation of public failure or worse.

8. *Wilmington Morning Star*, November 15, 1910.

9. Jane, *Jane's All the World's Aircraft*, 357; *Gouverneur v. Harriman Motor Works*, New Hanover Superior Court Minutes, 77.

10. Jane, *Jane's All the World's Aircraft*, 357.

11. *Raleigh News and Observer*, August 11, 1963.

12. *New York World*, May 12, 1912.

13. Ibid.

14. *New York Evening Post*, May 10, 1912.

15. *New York Evening Post*, May 3, 1910; *Raleigh News and Observer*, August 31, 1963.

16. *Wilmington Morning Star*, November 12, 1913; *Wilmington Morning Herald*, February 13, 1914.

17. *Wilmington Morning Star*, March 10, 31, 1917.

18. *Wilmington Evening Dispatch*, January 11, 1911.

19. Mackworth-Praed, *Aviation*, 111, 127.

20. Jane, *Jane's All the World's Aircraft*, 327; *Popular Mechanics*, June 1910, plate 1; *Kinston Free Press*, February 27, 1911, October 30, 1927, March 5, 1928, December 26, 1934.

21. *Kinston Free Press*, March 2–3, 1911; *Raleigh News and Observer*, March 10, 1911.

22. *Kinston Free Press*, March 7, 13, 1911.

23. *Wilson Times*, March 10, 1911; *Charlotte Evening Chronicle*, July 8, 1912.

24. *Charlotte Observer*, May 24, 1910, July 8, 1912; *Charlotte Evening Chronicle*, July 8, 1912.

25. Lowe, "Johnny Crowell."

26. *Winston-Salem Journal*, March 14, April 9, May 2–3, December 8, 1911, March 14, 1912.

27. Dula, *Durham and Her People*, 187; *Durham Morning Herald*, June 30, 1911.

28. *Durham Morning Herald*, June 18, 20, 30, August 17, October 9, 1911; *Charlotte Observer*, June 30, 1911; *Greensboro Daily News*, July 4, 1911; *Durham Recorder*, July 4, 1911; *Raleigh News and Observer*, August 24, 1911; *Statesville Sentinel*, August 31, 1911.

29. Mackworth-Praed, *Aviation*, 158.

30. *Charlotte Observer*, March 18, 1910; *Asheville Citizen*, February 2, 9, March 23, 27, 1910.

31. *Asheville Gazette-News*, February 5, March 4, June 23, 1910.

32. *Asheville Gazette-News*, June 17, 1910.

33. *Raleigh Times*, October 18, 1910.

34. *Greensboro Telegram*, July 26, 1911; *Greensboro Daily News*, May 26, June 4, July 17, 1915; Vecsey and Dade, *Getting off the Ground*, 79–80; Undated clipping, Forrest E. Wysong Biographical File.

35. *Greensboro Daily News*, October 1–2, 28, 1911, August 1, 1912, July 20, 1913; *Raleigh News and Observer*, October 4, 1911. For more on Andrews's flying career, see Chapter 8.

36. *Who Was Who in America, 1897–1942*, 586; *Greensboro Daily News*, August 1, 1912; *Charlotte Observer*, June 12, 1932.

37. *Asheville Citizen*, January 2, April 4, 1910.

38. *Asheville Citizen*, July 28, 1910, November 22, 1911, January 2, 1912; *Winston-Salem Journal*, January 4, 1910; *Wilson Times*, January 3, 1911.

39. Mrs. Gloria Lenoir Sodergren (daughter of James S. Spainhour), telephone interview by author, February 17, 1997. According to Alumni Office records, Spainhour attended the North Carolina College of Agriculture and Mechanic Arts as a special student (1897–99) but did not earn a degree.

40. *Pittsburgh Dispatch*, September 22, 1911; Trimble, *High Frontier*, 78.

41. *Pittsburgh Dispatch*, September 22, 1911; *Official Gazette of the United States Patent Office*, 1910, p. 202, 1914, pp. 249–51; Trimble, *High Frontier*, 78.

42. *Pittsburgh Dispatch*, September 22, 1911; Mrs. Sodergren, telephone interview by author.

43. Mrs. Sodergren, telephone interview by author; *Morganton Herald*, October 5, 1911; Mrs. Sodergren, telephone interview by author.

44. Stearns, "Flying Wing," 234–35; Harold E. Morehouse, "Dr. William W. Christmas," manuscript dated December 28, 1970, Christmas Biographical File.

45. Morehouse, "Dr. William W. Christmas."

46. Ibid.

47. Stearns, "Flying Wing," 235; E. Morehouse, rough notes, Christmas Biographical File.

48. Albert F. Zahm to Christmas, November 11, 1942, Christmas Biographical File; Villard, *Contact*, 215, 247.

49. Whitehouse, *Early Birds*, 106; *Baltimore Sun*, March 10, 1949, clipping in Robert Ions Biographical File; Morehouse, "Dr. William W. Christmas." Morehouse, apparently on the testimony of Christmas, lists him as the third American to fly. The Ions eulogy makes it appear that Ions (not Christmas) flew in late 1908 after Baldwin, Glenn Curtiss, J. A. D. McCurdy, and, probably, Samuel F. Cody. The Wrights were the only American predecessors of these airmen. The Christmas Aeroplane Company was incorporated in Washington, D.C., in the fall of 1910, with Christmas as president and Ions as vice president.

50. Stearns, "Flying Wing," 234–35; Whitehouse, *Early Birds*, 106; Edward Young to Harold E. Morehouse, October 8, 1929, Christmas Biographical File; *Charlotte Observer*, July 1, 1910.

51. Unidentified clipping, October 16, 1911, Christmas Biographical File.

52. *United States Patent Office Gazette*, 1910, p. 202, 1914, pp. 129–30; Shamburger and Christy, *Command the Horizon*, 63; Whitehouse, *Early Birds*, 106–7; Congressional Reference Division to Hon. Lee Hamilton, March 10, 1972, typescript, Christmas Biographical File. The letter to Hamilton says that the Reference Division searched U.S. Claims Reports for 1918–23 but found no record of Christmas's claim. The Report of the Secretary of War to the President, 1924, cites consideration "given to 20 cases filed with the War Department, alleging infringement of patents." Two claims were settled for a total of $103,500 but were not identified in the report. See Christmas Biographical File.

53. Unidentified clipping, April 15, 1912, Christmas Biographical File.

54. Unidentified clippings, May 21, June 1, 8, 1912, Christmas Biographical File; *New York Times*, May 12, 1912.

55. Beach, "New York Aero Show," 483.

56. Ibid.

57. *New York World*, May 16, 1921; *New York Times*, May 22, 1912.

58. Unidentified clippings, May 8, 27, 1912, Christmas Biographical File; *New York Times*, June 7, 1912.

59. *New York Times*, June 7, 1912.

60. Unidentified clipping, June 8, 1912, Christmas Biographical File. A clipping from *Linns Stamp News*, April 26, 1976, in the Christmas Biographical File, cites poor weather and precarious finances of the Christmas company as reasons for the cancellation.

61. Unidentified clipping, June 8, 1912, Christmas Biographical File.

62. Author's telephone conversations with Alumni Offices at George Washington University, February 23, 2000, and Johns Hopkins University, June 6, 1999 (Christmas's graduation from George Washington at age forty may mean that he received his inheritance only after the turn of the century); Christmas, "A Brief Authoritative Digest of the Most Important Works Done in Aviation by Dr. William W. Christmas," unsigned manuscript in the style and perspective of Christmas himself, Christmas Biographical File.

63. Burnelli with Herndon, "Wackiest Plane," 56.

64. *Washington (D.C.) Post*, April 15, 1960.

65. Stearns, "Flying Wing," 235.

66. *New York Evening Post*, March 10, 1912.

67. Beach, "New York Aero Show," 483; *New York World*, May 10, 1912.

68. *Raleigh News and Observer*, August 1, 1963.

69. *New York World*, May 10, 1912; *Wilmington Morning Star*, March 30, 1917.

CHAPTER EIGHT

1. Undated note from Thomas A. Allison (grandnephew of Raymond Allison) to author; *Austin (Tex.) Hornet*, a training camp newspaper at the School of Aeronautics, December 21, 1918. Upon graduation, Allison married Mary Rasel. Cornell University Alumni Office staff, telephone conversation with author, March 24, 1999.

2. Undated clipping [1910] regarding Allison as state agent for K.R.I.T automobiles, in possession of Thomas A. Allison.

3. *Daily Charlotte Observer*, September 15, 1910.

4. *Charlotte News*, October 21, 1910.

5. *Daily Charlotte Observer*, September 15, 1910.

6. *Charlotte News*, October 24, 28, 31, 1910.

7. *Charlotte Evening Chronicle*, November 8, 1910.

8. Ibid., November 10, 1910; *Daily Charlotte Observer*, November 10, 1910.

9. *Charlotte Chronicle*, November 8, September 9, 1910; *Statesville Sentinel*, September 15, 1910.

10. *Charlotte Chronicle*, November 9, 1910; *Daily Charlotte Observer*, November 9, 1910.

11. *Charlotte News*, November 10, 1910.

12. Ibid.

13. *Wilmington Morning Star*, November 26, 1910; Agreement between R. V. Allison and F. E. Schoonmaker, for the Industrial Motor Vehicle Company, July 7, 1911, regarding payments to Allison until the first ten trucks were sold, and Undated [1911] clipping regarding trucks to be built by the Industrial Motor Vehicle Company, both in possession of Thomas A. Allison.

14. *Charlotte News*, November 11–12, 1910.

15. Liberatore, *Helicopters before Helicopters*, 32.

16. *Asheville Citizen*, November 23, 1910; *Statesville Landmark*, December 1, 1910.

17. *Asheville Citizen*, November 23, 1910.

18. Ibid.

19. West, "Insecure Icarus," *This Proud Land*, 190. West "grew up" with this tale and did not recall where he first heard it. I am indebted to him for permission to quote part of his poem.

20. Mansfield, *Skylark*, 20–24; Crouch, *Bishop's Boys*, 428–32.

21. *Raleigh News and Observer*, November 15, 1910.

22. *Raleigh News and Observer*, October 28, 30, November 4, 1910.

23. *Raleigh News and Observer*, November 2, 1910.

24. *Southern Pines Tourist*, November 25, 1910.

25. *Wilmington Morning Sun*, November 29, 1910; *Raleigh News and Observer*, November 20, 1910; *Daily Charlotte Observer*, December 3, 1910.

26. Mansfield, *Skylark*, 24.

27. *Wilmington Morning Star*, January 6, March 11–12, 1911.

28. *Pinehurst Outlook*, March 11, 25, April 8, 1911.

29. *Pinehurst Outlook*, April 15, 1911.

30. Mansfield, *Skylark*, 20–23.

31. *Wilmington Morning Star*, April 19, 1911; *Wilmington Evening Dispatch*, October 28, 1910; *Salisbury Carolina Watchman*, November 22, 1911.

32. *Raleigh News and Observer*, May 18, 1940; *Durham Morning Herald*, May 4, 1911.

33. *Greensboro Daily News*, April 6, 1911.

34. *Raleigh News and Observer*, October 22, 17, 1911; *Greensboro Daily News*, October 5, 1911, April 25, 1912.

35. See, e.g., *Charlotte Observer*, October 23, 1907, and Cotten, *Light and Air*, 21.

36. *Greensboro Daily News*, October 10, 1911; *Raleigh News and Observer*, December 14, 1913; *Charlotte Chronicle*, February 11, 1911.

37. *Raleigh News and Observer*, October 19, 22, 27, November 1–2, 1911; *Winston-Salem Journal*, November 29–30, 1911; *Daily Charlotte Observer*, October 18, 1911; *Asheville Citizen*, November 4–5, 1911; *Elizabeth City Advance*, October 25, 1911.

38. *Wilkesboro Chronicle*, September 20, October 6, 1911; *Wilmington Morning Sun*, December 31, 1911, January 3, 1912.

39. *Winston-Salem Journal*, September 3, 1911.

40. *Raleigh News and Observer*, October 4, 1911; *Greensboro Daily News*, July 20, 1913; *Daily Charlotte Observer*, May 21, 1912.

41. *Raleigh News and Observer*, October 4, 1911; *Greensboro Daily News*, October 8, 17, 1911; *Charlotte Observer*, June 12, 1932.

42. *Greensboro Daily News*, October 17, 1911.

43. *Greensboro Daily News*, August 1, 1912.

44. *Greensboro Daily News*, August 1, September 20, 1912; *Charlotte Observer*, June 12, 1932.

45. *Charlotte Observer*, August 2, 1911, July 20, 1913.

46. See, e.g., *Daily Charlotte Observer*, July 2, 1912; *Raleigh News and Observer*, July 5, 1913; *Greensboro Daily News*, October 11, 1913; and *Winston-Salem Journal*, September 14, 1913.

47. *Wilmington Morning Star*, January 24, 1915.

48. Vecsey and Dade, *Getting off the Ground*, 30–31.

49. Ibid., 32.

50. Ibid., 32–33.

51. *Asheville Citizen*, May 17, 1911.

52. Vecsey and Dade, *Getting off the Ground*, 32–33; Tiny Broadwick's 1952 application to join the Early Birds, Tiny Broadwick Biographical File.

53. Vecsey and Dade, *Getting off the Ground*, 33–34.

54. Ibid., 34–35.

55. Ibid.

56. Ibid., 35–36.

57. Ibid.

58. Ibid., 36–37; *Greenville Daily Reflector*, September 6, 1978.

59. *Charlotte Observer*, May 8, September 7, December 28, 1910; Scott, *Shoulders of Giants*, 245.

60. *Charlotte Observer*, September 7, November 3, 1910.

61. *Charlotte Observer*, December 9, 28, 1910.

62. *Charlotte Observer*, June 4, 1911.

63. *Greensboro Daily News*, October 3, 1912.

64. *Wilmington Morning Star*, January 4, 1911.

65. *Salisbury Evening Post*, November 28, 1910.

66. *Charlotte Observer*, December 12, 1912; *Raleigh News and Observer*, November 19, 1910.

67. *Wilmington Morning Star*, December 21, 1910.

68. *Greensboro Daily News*, September 10, 1912.

69. *Raleigh News and Observer*, November 19, 1910; *Salisbury Carolina Watchman*, August 10, 1910.

70. *Daily Charlotte Observer*, July 2, 1910.

CHAPTER NINE

1. *Greensboro Telegram*, July 26, 1911; "Forrest Wysong: Our New President," *Chirp*. *Chirp* is the magazine of the Early Birds organization.

2. Forrest Wysong: Our New President"; Vecsey and Dade, *Getting off the Ground*, 79.

3. *Greensboro Telegram*, July 26, 1911; "Forrest Wysong: Our New President."

4. Vecsey and Dade, *Getting off the Ground*, 78–79.

5. Ibid., 79.

6. "Forrest Wysong: Our New President," 3.

7. "Forrest E. Wysong," unidentified typescript of undated interview with Wysong, 2–4, Wysong Biographical File; Vecsey and Dade, *Getting off the Ground*, 79.

8. Vecsey and Dade, *Getting off the Ground*, 79.

9. Undated clipping, endorsed "probably April 5, 1915," *Greensboro Record*, Wysong Biographical File.

10. Vecsey and Dade, *Getting off the Ground*, 79–80.

11. *Greensboro Daily News*, June 14, 26, July 17, 1915.

12. *Greensboro Daily News*, May 26, June 14, July 17, 1915.

13. "Forrest Wysong: Our New President," 3–5; Forrest E. Wysong to Charles F. Willard, July 10, 1961, typescript, Wysong Biographical File.

14. Mackworth-Praed, *Aviation*, 156.

15. Ibid., 171; Longstreet, *Canvas Falcons*, 31–68; Norman, *Great Air War*, 1–7.

16. *Wilmington Morning Herald*, March 15, 1917.

17. Hall, "William D. Polite"; *Wilmington Morning Herald*, March 15, 1917, August 31, 1911.

18. *Official Gazette of the United States Patent Office*, 1917, 197–98.

19. *Wilmington Morning Herald*, October 22, 1915.

20. *New York Times*, November 12, December 5, 1915.

21. *New York Times*, December 5, 1915.

22. Ibid.

23. Harold E. Morehouse, "Dr. William W. Christmas," manuscript dated December 28, 1970, Christmas Biographical File.

24. Advertisement in *Flying* 4 (September 1915): 21.

25. *Raleigh News and Observer*, November 29, 1917, February 9, 1918; April 12, 1916.

26. "Daniels, Josephus," Powell, *Dictionary of North Carolina Biography*, s.v.

27. Daniels, *Wilson Era*, 1:290; *Raleigh News and Observer*, March 1, 1914.

28. Daniels, *Wilson Era*, 1:290, 292; *Raleigh News and Observer*, May 22, 1913.

29. *Raleigh News and Observer*, December 20, 1928.

30. Daniels, *Wilson Era*, 1:293, 293–96; Daniels, *Our Navy at War*, 228.

31. *Raleigh News and Observer*, November 21, 1913.

32. Ibid.

33. Ibid.

34. Ibid.

35. Tillman, *Man Unafraid*, 74–77.

36. Maurer, "First Aero Squadron," 207–8; Tillman, *Man Unafraid*, 172–73.

37. *Raleigh Times*, October 12, 1910; *Clayton Chronicle*, October 11, 1917; *Raleigh News and Observer*, October 8, 1917.

38. *Clayton News*, November 7, 1979.

39. *San Diego Tribune*, November 26, 1913; "Army Airplane Receipts."

40. Unidentified clipping, Ellington File.

41. *San Diego Union*, November 25, 1913; *Wilmington Morning Herald*, November 25, 1913; "Mortality in Army Aviation."

42. *San Diego Union*, November 17, 24, 1913; *Clayton Chronicle*, October 11, 1917.

43. *San Diego Chronicle*, November 25, 1913.

44. "The Men and the Machines," 174; Foulois, "Early Flying Experiences," 172.

45. Bonney, "Chiefs of the Army Air Force," 130.

46. "Army Airplane Receipts"; Norman, *Great Air War*, 20–21.

47. Unidentified clipping, December 1913, Ellington File.

48. Tillman, *Man Unafraid*, 199–201.

49. Clipping from *Greensboro Record*, 1935, copy furnished to author by Charles E. Moore.

50. Ibid.; *Greensboro Record*, March 30, 1929.

51. Norman, *Great Air War*, 3; Tillman, *Man Unafraid*, 195; *Raleigh News and Observer*, February 27, 1917.

52. *Raleigh News and Observer*, June 6, 1915.

53. *Raleigh News and Observer*, March 31, 1929.

54. *Warrenton Warren Record*, July 23, 1917, February 8, 1917.

55. *Wilmington Morning Star*, September 6, 1915, March 1, 1917.

56. "Kiffin Yates Rockwell," first page of unidentified three-page typescript, Kiffin Y. Rockwell Biographical File; *Asheville Citizen-Times*, September 21, 1952.

57. Thomas Wolfe, *Look Homeward, Angel*, 287; *Asheville Citizen-Times*, September 21, 1952. In Wolfe's novel the Rockwell brothers are referred to as Paul and Clifton Wheeler.

58. *Asheville Citizen-Times*, September 21, 1952.

59. Ibid.; *Asheville Citizen-Times*, June 9, 1964; Paul A. Rockwell Clipping File, North Carolina Collection, box 67; Parsons, *I Flew with the Lafayette Escadrille*, 155–56.

60. *Raleigh News and Observer*, August 29, 1920.

61. Norman, *Great Air War*, 257–58, 253.

62. Rockwell, *American Fighters*, 187–88.

63. *Raleigh News and Observer*, April 30, 1922, February 17, 1971; *Southern Pines Pilot*, February 3, 1931.

64. *Raleigh News and Observer*, April 30, 1922; McConnell, *Flying for France*, 41–53.

65. McConnell, *Flying for France*, 26–27; Kiffin Rockwell to Mrs. Loula Ayres

Rockwell, May 23, 1917, Military Collections, World War I Papers, box 67; Norman, *Great Air War*, 272–73.

66. Norman, *Great Air War*, 275.

67. Ibid., 275–76; Parsons, *I Flew with the Lafayette Escadrille*, 3–6.

68. Parsons, *I Flew with the Lafayette Escadrille*, 4–6; Rockwell, *American Fighters*, 188–89.

69. Parsons, *I Flew with the Lafayette Escadrille*, 6–7; Kiffin Rockwell to Mrs. Rockwell, May 23, 1917, Military Collections, World War I Papers, box 67.

70. Norman, *Great Air War*, 277–78.

71. Rockwell, "Writings of the American Pilots"; Norman, *Great Air War*, 288, 308.

72. Norman, *Great Air War*, 533–35.

73. Parsons, *I Flew with the Lafayette Escadrille*, 157–58; Rockwell, *American Fighters*, 190.

74. Parsons, *I Flew with the Lafayette Escadrille*, 157–58.

75. Ibid.

76. Norman, *Great Air War*, 284–85.

77. Ibid., 288–89; Rockwell, *American Fighters*, 195.

78. Rockwell, *American Fighters*, 196–98; *Durham Herald-Sun*, April 20, 1941.

79. Parsons, *I Flew with the Lafayette Escadrille*, 160–61; Rockwell, *American Fighters*, 197; McConnell, *Flying for France*, 138.

80. *Asheville Citizen-Times*, September 21, 1952.

81. *Raleigh News and Observer*, April 30, 1922; Rockwell, *American Fighters*, 195; McConnell, *Flying for France*, 76; *Carthage Moore County News*, April 5, 1917.

82. *Raleigh News and Observer*, April 30, 1922; *Carthage Moore County News*, April 5, 1917; Rockwell, *American Fighters*, 194–95.

83. McConnell, *Flying for France*, 54–55.

84. *Carthage Moore County News*, April 5, 1917; Parsons, *I Flew with the Lafayette Escadrille*, 253.

85. Walter Brown, *An American for Lafayette*, 164–66, xxiii–xxiv, 171, 208; Parsons, *I Flew with the Lafayette Escadrille*, 246–47.

86. Parsons, *I Flew with the Lafayette Escadrille*, 246–47.

87. *Southern Pines Pilot*, February 17, 1971; *Raleigh News and Observer*, April 30, 1922.

88. *Raleigh News and Observer*, April 30, 1922; Paul A. Rockwell to McDaniel Lewis, January 31, 1971, copy in Paul A. Rockwell Clipping File, North Carolina Collection.

CHAPTER TEN

1. Norman, *Great Air War*, 2–3.

2. *Wilmington Morning Star*, September 12, 1913, January 3, June 20, 1918, May

21, 1938; *Charlotte Observer*, June 21, 1918; Norman, *Great Air War*, 306–7; Mattox, "Chariots of Wrath," 297.

3. *Charlotte Observer*, June 21, 1916; *Raleigh News and Observer*, May 18, 1940.

4. Mattox, "Chariots of Wrath," 298; *Raleigh News and Observer*, June 16, 1918.

5. *Wilmington Morning Star*, June 20, 1918; Mattox, "Chariots of Wrath," 299.

6. Mattox, "Chariots of Wrath," 299; *Alumni Magazine*, North Carolina State College, 1918, 4.

7. Mattox, "Chariots of Wrath," 299–300; Baugham to his mother, May 4, July 7, 1917, and "James Henry Baugham," Baugham Papers; *Alumni Magazine*, North Carolina State College, 1918, 4; Mattox, "Chariots of Wrath," 300.

8. Baugham to his mother, February 25, 1918; *Alumni Magazine*, North Carolina State College, 1919, 4.

9. *Raleigh Times*, July 4, 1918.

10. Ibid.

11. *Charlotte Observer*, August 17, 1920; Mattox, "Chariots of Wrath," 301–2.

12. *Warrenton Warren Record*, October 22, 1917.

13. Ibid.; Smith, *First Across!*, 14.

14. *Warrenton Warren Record*, October 22, 25, 1917.

15. *Warrenton Warren Record*, October 22, 1917.

16. For Marshburn and Winston, see *Raleigh News and Observer*, May 23, 1927; for the Rorisons, see *Wilmington Morning Star*, September 12, 1917; for Woody, see *Who's Who in Aviation*, 118; for McLean and Renn, see Chap. 9 above; for Bishop, see *Raleigh News and Observer*, November 6, 1917.

17. See, e.g., *Alumni News*, North Carolina State College 15 (1918): 8.

18. *Raleigh News and Observer*, November 30, 1917.

19. *Warrenton Warren Record*, September 7, 1917.

20. *Warrenton Warren Record*, November 14, 1918.

21. *Alumni Review*, University of North Carolina 5 (1918): 178.

22. *Charlotte Observer*, September 4, 1927.

23. *Asheville Citizen*, May 1, 1932.

24. David A. Davis to his mother, February 16, 1918 (box 6), and Elbert Ezra Wilson to his mother, November 2, 1917 (box 72), Military Collections, World War I Papers.

25. *Warrenton Warren Record*, April 23, 1918.

26. *Smithfield Herald*, November 11, 1918; *Stars and Stripes*, March 22, 1918.

27. *Raleigh News and Observer*, July 17, 1918; *Warrenton Warren Record*, September 7, 1917.

28. *Raleigh News and Observer*, January 23, 1918; Palmer to his mother, quoted in *Warrenton Warren Record*, April 3, 1918.

29. Cobb to his mother, January 12, 1918, Military Collections, World War I Papers, box 12; *Raleigh News and Observer*, January 23, 1918.

30. *Charlotte Observer*, October 13, 1940.

31. *Raleigh News and Observer*, July 7, 1918.

32. For Robbins's death, see *Raleigh Times*, July 6, 1918; for Wilford's, see *Raleigh News and Observer*, July 14, 1918.

33. *Raleigh News and Observer*, July 8, 1918.

34. Ibid.

35. For Sykes, see *Charlotte Observer*, September 14, 1918; Loughran, Military Collections, World War I Papers, box 1; for Olive, see "Old Gold and Black," Wake Forest College, March 1, 1920, Baptist Collection, Wake Forest University Library.

36. *Charlotte Observer*, August 21, 1918; *Greensboro Daily News*, August 21, 1918; *Asheville Citizen*, October 20, 1918.

37. *Asheville Citizen*, October 20, 1918.

38. *Charlotte Observer*, October 3, 1940.

39. Ibid.

40. *Wilson Daily Times*, August 28, 1918.

41. *Reidsville Review*, October 15, 1918.

42. Ibid.

43. Citations, Military Collections, World War I Papers, box 7.

44. *Reidsville Review*, February 4, 1919; *Charlotte Observer*, February 7, 1919; *Asheville Citizen*, July 17, 1927; Compiled Individual Service Records, box 7; *Alumni News*, North Carolina State College, 19 (December 1, 1918): 3.

45. *Reidsville Review*, February 4, 1919.

46. *Wilmington Morning Star*, September 12, 1917, October 17, 1919; *Charlotte Observer*, September 19, 1932; Military Collections, World War I Papers, box 9.

47. Military Collections, World War I Papers, box 9.

48. Pou to his father, September 16, 1918, Pou Papers; *Raleigh News and Observer*, July 14, 1918; *Smithfield Herald*, November 3, 1918.

49. *Raleigh News and Observer*, July 7, 1919; *Alumni News*, North Carolina State College 18 (1919): 2.

50. *Aircraft Yearbook*, 287–93; Military Collections, World War I Papers, box 59.

51. Mansfield, *Skylark*, 7–8.

52. *Raleigh Times*, February 26, March 18, 1918.

53. *Raleigh Times*, March 18, 27, June 21, July 24, 1918.

54. *Raleigh Times*, March 27, June 21, 1918.

55. Mansfield, *Skylark*, 48–94.

56. *Raleigh Times*, July 24, 30, August 18, 27, 1918; Mansfield, *Skylark*, 95–96.

57. Mansfield, *Skylark*, 96–98.

58. Ibid., 98–99; *Raleigh News and Observer*, December 20, 1918.

59. *Raleigh News and Observer*, December 8, 1915; *Charlotte Observer*, August 24, 1918.
60. *Raleigh News and Observer*, August 24, 1918, March 3, 1919; *Charlotte Observer*, August 24, 1918.
61. *Beaufort News*, July 19, 1920.
62. *Raleigh News and Observer*, March 2, 1917, March 31, 1920.
63. *Beaufort News*, July 29, 1920.
64. Winters, "Life-Saving Airplanes."
65. Burnelli with Herndon, "Wackiest Plane," 74, 76.
66. Ibid., 72, 76.
67. Ibid., 72.
68. Ibid., 56, 72.
69. Ibid., 75–76.
70. Ibid., 76–77.
71. Van Vliet, "Inside Story," 37.
72. Burnelli with Herndon, "Wackiest Plane," 78; Lt. E. E. Hall to General Development Co. of Connecticut, February 8, 1930, Christmas Biographical File.
73. Burnelli with Herndon, "Wackiest Plane," 78.
74. Bellefleur, "Last Early Aeroplane?," 42. I am indebted to Professor Bellefleur for a copy of his manuscript.
75. Burnelli with Herndon, "Wackiest Plane," 78–79.

CHAPTER ELEVEN
1. *Raleigh News and Observer*, January 9, April 8, 1919.
2. *Fayetteville Observer*, February 1, June 16, 1919; *Charlotte Observer*, May 1, 1919; *Raleigh News and Observer*, May 1, 8, 1921, April 30, 1927.
3. *Fayetteville Observer*, July 5, 1919, July 5, 1923, August 12, 1924; *Raleigh News and Observer*, April 30, 1927.
4. *Raleigh News and Observer*, May 17, 1942; *Asheville Citizen*, June 9, 1966; "Paul Rockwell," Paul A. Rockwell Clipping File.
5. Karolevitz and Fenn, *Flight of Eagles*, 58.
6. Ibid., 78–79, 121, 129, 159, 186, 214.
7. *Who's Who in Aviation*, 89, 91, 191.
8. *Raleigh News and Observer*, October 22, 1918, May 22, 1927.
9. On Privette and Bishop, see *Raleigh News and Observer*, June 13, 1919; on Leazar, see *Charlotte Observer*, July 20, 1919; on Westall, see *Asheville Citizen*, September 3, 1949; on Woody and Pennington, see *Wilmington Morning Star*, January 21, 1976; on McLean, see C. E. "Eddie" Moore to the author, September 26, 1999. At this writing Moore, a grandnephew of Hubert McLean, is operator of McLean Brothers Airport.

10. *Statesville Landmark*, July 16, August 13, 1918; *Raleigh News and Observer*, June 15, 1919.

11. *Fort Sill (Okla.) Trench and Camp*, May 2, 1919, photocopies. His Commission of June 17, 1919, and Honorable Discharge of May 29, 1919, were furnished to the author by Thomas A. Allison, grandnephew of Raymond Allison.

12. William A. Allison to Raymond Allison, May 31, 1919, photocopy furnished to author by Thomas A. Allison; Thomas A. Allison, telephone interview by author, February 17, 1996.

13. Thomas A. Allison, telephone interview by author, October 19, 1995.

14. *Statesville Sentinel*, April 17, 1922; Alumni Office staff, telephone interview by author, Cornell University, September 19, 1996.

15. For Cooley, see Powell, *Biographical Dictionary of North Carolina*, s.v.; for Currie, see *Alumni News*, North Carolina State College 17 (1919): 4; for Carmichael, see *Raleigh News and Observer*, August 13, 1918, and Powell, *Dictionary of North Carolina Biography*, s.v.; for Morehead, see *Alumni Review*, University of North Carolina 6 (1918): 217; for Stephenson, see *Alumni Review*, University of North Carolina 7 (1919), 217, and Powell, *Dictionary of North Carolina Biography*, s.v.

16. Norris, "Flying Parson," 18; *Raleigh News and Observer*, October 29, 1918.

17. Rooks, "Memories of Belvin W. Maynard," photocopy in author's possession.

18. *Dell School Catalogue*, 6, 10, 15; O'Kelly, "Notes on a Conversation with Mary Kate Allen," 7, O'Kelly Papers.

19. O'Kelly, "Notes on a Conversation with Mary Kate Allen," 1, 7; Rooks, "Memories of Belvin W. Maynard"; Unidentified newspaper clipping, 1919, furnished to author by Mrs. Caralee A. Rooks.

20. *Raleigh News and Observer*, September 8, 14, 1923, September 8, 1922; O'Kelly, "Notes on a Conversation with Mary Kate Allen," 7; Rooks, "Essie Goodson Maynard," copy in author's possession.

21. *Raleigh News and Observer*, September 8, 1922, April 17, 1919.

22. *Raleigh News and Observer*, April 17, 1919.

23. Ibid.

24. Bowers, "Aviation on Trial," 90; Rooks, "Essie Goodson Maynard"; Arthur, "Flying Parson," 12; *New York Times*, August 24, 1919.

25. "Toronto–New York Handicap Race," 20; *New York Times*, August 24–26, 1919.

26. *New York Times*, August 26, 1919; "Toronto–New York Handicap Race," 20.

27. *New York Times*, August 30, 1919; *Raleigh News and Observer*, September 3, 1919.

28. *Raleigh News and Observer*, September 8, 1922; "Double Transcontinental Aerial Derby," 29.

29. *Raleigh News and Observer*, October 24, 1919.

30. Ibid. This was Maynard's own account of the race, extending over subsequent installments.

31. *Raleigh News and Observer*, October 25, 1919.

32. *Raleigh News and Observer*, October 25–26, 1919.

33. *Raleigh News and Observer*, October 26–27, 1919.

34. *Raleigh News and Observer*, October 28, 1919; Bowers, "Aviation on Trial," 92.

35. *New York Times*, October 12, 1919; "Double Transcontinental Aerial Derby," 6–11.

36. *Raleigh News and Observer*, October 28, 1919.

37. *Raleigh News and Observer*, October 29, 1919.

38. *Raleigh News and Observer*, October 30, 1919.

39. Ibid.

40. Ibid.

41. *Raleigh News and Observer*, October 31, 1919.

42. *New York Times*, October 19, 1919; "Double Transcontinental Aerial Derby," 6.

43. *New York Times*, October 19, 1919; *Raleigh News and Observer*, October 29, 1919.

44. Bowers, "Aviation on Trial," 99.

45. *New York Herald Tribune*, October 19, November 2, 1919.

46. Bowers, "Aviation on Trial," 96, 98.

47. Ibid., 99–100.

48. *Raleigh News and Observer*, October 20, 1919, November 2, 1919.

49. *Raleigh News and Observer*, November 2, 1919; Gresham, "Editor's Portfolio," 140.

50. *Raleigh News and Observer*, November 4, 1919.

51. *Raleigh News and Observer*, November 7, 1919.

52. *Raleigh News and Observer*, April 6–7, May 4, 7, 11, 29, June 2, 6, 12, July 4, December 6, 8, 1919; *New York Herald Tribune*, November 26, 1919.

53. *New York Herald Tribune*, December 12, 19, 1919.

54. *New York Herald Tribune*, December 9, 1919.

55. *New York Herald Tribune*, December 10, 1919; *New York Times*, December 10, 12, 1919.

56. *New York Herald Tribune*, December 10, 1919; *New York Times*, December 10–11, 1919.

57. *New York Times*, December 12–13, 17, 1919; *Raleigh News and Observer*, December 19, 1919.

58. *New York Herald Tribune*, December 22, 1919; *New York Times*, December 22, 1919.

59. Bowers, "Aviation on Trial," 98.

60. *New York Times*, February 12, May 9, 1920; *Raleigh News and Observer*, February 10, 15, 1919.

61. *Raleigh News and Observer*, February 22, 29, March 28, 30–31, April 4, 9, 1920.

62. *Raleigh News and Observer*, December 13, 1953.

63. *Raleigh News and Observer*, May 8, 1920, September 8, 1922; Rooks, "Memories of Belvin W. Maynard."

64. *Raleigh News and Observer*, September 8, 1922; *Rutland (Vt.) Daily Herald*, September 7–8, 1922.

65. *Raleigh News and Observer*, September 8, 1922.

66. Ibid.

67. *Raleigh News and Observer*, November 2, 1919.

68. *Raleigh News and Observer*, December 13, 1953.

CHAPTER TWELVE

1. Smith, *First Across!*; Mason and Windrow, *Air Facts and Feats*, 124; *Aircraft Yearbook*, 258–59.

2. Levine, *Mitchell*, 231–47.

3. Ibid., 281–88; MacNeill, *Hatterasman*, 241.

4. MacNeill, *Hatterasman*, 241–42.

5. Levine, *Mitchell*, 215.

6. Ibid., 288–89; MacNeill, *Hatterasman*, 242.

7. Levine, *Mitchell*.

8. *Aircraft Yearbook*, 323; "Fields and Sites"; "Landing Fields in the United States."

9. *Wilmington Morning Star*, September 13, 1921.

10. *Raleigh News and Observer*, February 1, 1920.

11. Ibid.

12. *Raleigh News and Observer*, April 13, 24, 1921, May 18, 1940.

13. *Raleigh News and Observer*, April 14, 27, 1921; Carr E. Booker, "Wright Brothers Plane in a British Museum. Why?," 4; Carr Booker Jr., interview by author, Raleigh, May 8, 1995.

14. *Wilmington Morning Star*, May 13, 1922.

15. Anonymous, *Flying* 9 (March 1920): 73.

16. Ibid., 9 (October 1920): 97.

17. *New York Times*, August 27, 1922.

18. Jane, *Jane's All the World's Aircraft*, 1922, 274b; *Greensboro Daily News*, November 26, 1929; Stearns, "Flying Wing," 234–35.

19. *Time*, April 20, 1960, 106; Whitehouse, *Early Birds*, 107. Several major newspapers had this to say on April 15, 1960: Christmas's wife, Mary Norris, died around 1950 (*Washington Post*); Christmas left one son, Whitney Christmas of Maryland (*New York Herald*); the son was named Frank (*New York Times*); Frank Christmas lived in Lawson, Md. (*New York Herald Tribune*). The small file of papers left by Christmas, now in the possession of his nephew, James

Donelson Christmas, includes a bequest to Mrs. Rosario de Munoz Morrison, his executor and probably the "luscious Latin beauty" mentioned by Burnelli with Herndon, "Wackiest Plane," 72. A published but unattributed article in the possession of James Donelson Christmas indicates that Mrs. Morrison died in November 1959. She was a native of Saville. Donelson Christmas, a pallbearer at his uncle's funeral, said that he "didn't have a dime" at the time of his death; interview with author, August 23, 1994. See also Stearns, "Flying Wing," 234–38. In short, Christmas's immediate family is no less mysterious than his own character and career.

20. Glines, *Roscoe Turner*, 19–20; *Raleigh News and Observer*, October 19, 1919. Runser became lost en route to Lenoir and spent a night on a farm near Morganton before reaching his destination.

21. *Winston-Salem Journal*, April 7, 1935.

22. Ibid.

23. Glines, *Roscoe Turner*, 22–23, 51; *Raleigh News and Observer*, October 20, 1919.

24. Glines, *Roscoe Turner*, 20, 27–29; *Raleigh News and Observer*, October 20, 22, 25, 27, 1919, November 11, 1919, January 26, February 28, 1920.

25. Glines, *Roscoe Turner*, 29; *Raleigh News and Observer*, January 26, February 28, 1920; *Charlotte Observer*, October 30, 1920.

26. Glines, *Roscoe Turner*, 38.

27. *Raleigh News and Observer*, August 19, 1926, June 21, July 21, 1927, November 27, 1919, May 18, 1940, December 12, 1953. I am indebted to Hal and David Stewart, Alton Stewart's sons, for data and photographs on his career.

28. Buffington, "Flying Life of Viola Gentry"; Brooks-Pazmany, *United States Women in Aviation*, 29; *Charlotte Observer*, December 22, 1928.

29. See also, e.g., W. C. Allen, *North Carolina History Stories*, Richmond: Johnson Publishing Co., 1927; Jule B. Warren, *North Carolina Yesterday and Today*, Raleigh: Edwards and Broughton Co., 1941; W. C. Allen, *The Story of the Old North State*, Raleigh: Dixie Press, 1942; Alex M. Arnett, *The Story of North Carolina*, Chapel Hill: University of North Carolina Press, 1942; Daniel Jay Whitener, *North Carolina History*, Oklahoma City: Harlow Publishing Corp., 1958; Blackwell P. Robinson, *The North Carolina Adventure*, Durham: Moore Publishing Co., 1968; and Thomas C. Parramore, *Carolina Quest*, Englewood Cliffs, N.J.: Prentice-Hall, 1978, and *North Carolina: The History of an American State*, Englewood Cliffs, N.J.: Prentice-Hall, 1983.

30. Rowe, *Discovering North Carolina*, 176–79; Lefler and Newsome, *North Carolina*, 681.

31. *Who's Who in Aviation*, 423; *Elizabeth City Independent*, December 17, 1937.

32. *Elizabeth City Tar Heel*, April 4, 1902, April 9, August 13, 1909, April 22, 1910.

33. Tate to Orville Wright, December 3, 1927, Wright Papers; W. J. Tate to

R. D. W. Conner, May 30, 1913, North Carolina Historical Commission Papers; Sidney A. Clark, *Cathedral France*, 199–200.

34. "Olds, Frederick Augustus," Powell, *Dictionary of North Carolina Biography*, s.v.; Olds, "Mother of the Airplane"; William J. Tate, "I Was Host . . . at Kitty Hawk," 192; *Warrenton Warren Record*, October 2, 1918; *Lexington Dispatch*, October 2, 1918.

35. Mrs. Willia Severn Raye (daughter of Irene Tate Severn), telephone interview by author, October 2, 1993; Harold E. Morehouse, "Edwin K. Jacquith: Flying Pioneers Biographies," three-page typescript, E. K. Jacquith Biographical File; *Elizabeth City Advance*, January 22, 26, 1917; *New York Times*, December 9, 1916. Jacquith owned the plane and was copilot.

36. Mrs. Raye, telephone interview by author.

37. Daniel Grady Tate (grandson of Dan Tate), telephone interview by author, September 4, 1993; Marden, "She Wore the World's First Wings."

38. *Raleigh News and Observer*, March 2, 1917.

39. *Wilmington Morning Star*, January 28–31, 1917; Reeves, *Southport (Smithwick)*, 255.

40. Mrs. Raye, telephone interview by author.

41. *Elizabeth City Daily Advance*, March 14–15, 18–19, 23, April 23, 1921; *Raleigh News and Observer*, March 17, 1921.

42. Tate to Orville Wright, March 19, 1922, Wright Papers; Olds, "Mother of the Airplane"; Orville Wright to Tate, May 19, 1922, Wright Papers.

43. Business card, Tate Biographical File.

44. *Raleigh News and Observer*, November 21, 1913; *Manteo Coastland Times*, December 11, 1953; *Chapel Hill Weekly*, February 8, 1927.

45. Tate to Fred Olds, May 15, 1927, Olds Papers; *Elizabeth City Independent*, April 22, 1927.

46. Tate to Orville Wright, May 24, 1927, Wright Papers; *Elizabeth City Independent*, April 22, 1927; Tate to Olds, May 15, 1927, Olds Papers.

47. *Charlotte Observer*, April 22, 1928; Tate, *Wings over Kill Devil*, 16–17.

48. *Charlotte Observer*, March 9, May 1, 16, 1926.

49. *Charlotte Observer*, April 22, 1928; *Elizabeth City Independent*, April 13, 1928; Tate to Orville Wright, March 18, May 6, 1928, Wright Papers.

50. *Elizabeth City Independent*, March 16, 1928.

51. *Elizabeth City Independent*, March 11, 30, December 4, 1928; *Charlotte Observer*, December 17, 1928. D. W. Hayman secured a lease on the Kill Devil Hill tract from his uncle, M. D. Hayman, and sold it the following year to Stick and his partner. *Elizabeth City Independent*, March 11, 1937.

52. Tate to Orville Wright, December 19, 1926, November 27, 1928, Wright Papers.

53. *Raleigh News and Observer*, June 12, 1928, December 18, 1940; *Elizabeth City*

*Independent*, May 4, 1928; "Captain William J. Tate"; *Chapel Hill Weekly*, February 3, 1933.

54. Tate, *Wings over Kill Devil*, 4; *Elizabeth City Advance*, December 13, 1937; Unidentified clipping, Tate Biographical File; "Captain William J. Tate," 6; Renstrom, *Wilbur and Orville Wright*, 94.

55. *Raleigh News and Observer*, January 20, 1937; *Charlotte Observer*, September 8, 1929.

56. See Tate, *Who's Who in Aviation*, 155; Tate, *Wings over Kill Devil*, 5.

57. *Raleigh News and Observer*, December 18, 1940; *Chapel Hill Weekly*, February 2, 1933; *Greensboro Daily News*, October 25, 1942.

58. Tate to Wright, May 1943, October 13, 30, 1948, May 1943, Wright Papers; *Raleigh News and Observer*, January 12, February 12, 1948.

59. Daniel Grady Tate, "W. J. Tate and Family," 419.

60. Tate to Wright, March 18, 1928, Wright Papers.

61. Orville Wright to Tate, April 3, 1928, ibid.

62. Fred Kelly to Lester D. Gardner, December 8, 19__, ibid.

63. *Charlotte Observer*, January 8, 1924, March 15, 1937, December 17, 1953.

64. Unidentified newspaper clipping in possession of Hal Stewart.

65. *Raleigh News and Observer*, April 4, 1918, February 13, 1919.

66. *Raleigh News and Observer*, April 4, 1919.

67. *New York Times*, September 22, 1922.

68. *Charlotte Observer*, February 9, 1930.

69. Ibid.

70. Ibid.

MANUSCRIPTS

Raymond Vance Allison Papers, in possession of Thomas A. Allison, Statesville

Alumni Biographical Files, Alumni Records Office, North Carolina State University, Raleigh

Daniel Asbury Biographical File, National Air and Space Museum, Smithsonian Institution, Washington, D.C.

Aviation File, Cape Fear Museum, Wilmington

Baptist Collection, Wake Forest University Library, Winston-Salem

James Baugham Papers, Manuscript Room, North Carolina State University Library, Raleigh

Carr E. Booker Papers, in possession of Roy L. Booker, Raleigh

Tiny Broadwick Biographical File, National Air and Space Museum, Smithsonian Institution, Washington, D.C.

William Wallace Whitney Christmas Biographical File, National Air and Space Museum, Smithsonian Institution, Washington, D.C.

William Wallace Whitney Christmas Papers, in possession of James Donelson Christmas, Upper Marlborough, Md.

J. T. Daniels Biographical File, National Air and Space Museum, Smithsonian Institution, Washington, D.C.

Dare County Record of Deeds, 1926–27, microfilm copy, North Carolina Division of Archives and History, Raleigh

Eric Ellington File, Ellington Memorial Library, Clayton, N.C.

John C. Emmerson Jr. Scrapbook, Sargeant Room, Kirn Library, Norfolk, Va.

Federal Population Census, Twelfth Census, 1900, Atlantic Township and Currituck County, microfilm copy, North Carolina Division of Archives and History, Raleigh

"Fields and Sites," file 686, Southeastern Department, Record Group 18 (Air Force), Southern Department, National Archives, Washington, D.C.

Bill Harris Audiotapes, Outer Banks History Center, Manteo

Robert Ions Biographical File, National Air and Space Museum, Smithsonian Institution, Washington, D.C.

E. K. Jacquith Biographical File, National Air and Space Museum, Smithsonian Institution, Washington, D.C.

Kramer Family Papers, East Carolina University Archives, Greenville

Belvin Womble Maynard Papers, in possession of Caralee Maynard Rooks, Mechanicsville, Va.

Hubert McLean Papers, in possession of Charles E. Moore, Gibsonville, N.C.

Al McSurely Papers, Chapel Hill

Military Collections, World War I Papers, 1903–22, Compiled Individual Service Records, North Carolina Division of Archives and History, Raleigh

New Hanover County Superior Court Minutes, 1911, microfilm copy, North Carolina Division of Archives and History, Raleigh

North Carolina Historical Commission Papers, Correspondence, box 11, 1913, North Carolina Division of Archives and History, Raleigh

Thomas O'Dwyer Diary, Southern Historical Collection, University of North Carolina, Chapel Hill

Sylvia Maynard O'Kelly Papers, Raleigh

Fred A. Olds Papers, North Carolina Division of Archives and History, Raleigh

William Luther Paul Papers, in possession of Charles L. Paul

William Luther Paul Papers, in possession of Grayden M. Paul

Edwin S. Pou Papers, Johnston County Library, Smithfield, North Carolina

Kiffin Y. Rockwell Biographical File, National Air and Space Museum, Smithsonian Institution, Washington, D.C.

Paul A. Rockwell Clipping File, North Carolina Collection, University of North Carolina Library at Chapel Hill

Joseph E. Shanks Papers, Sargeant Room, Kirn Library, Norfolk, Virginia

Orren Randolph Smith Scrapbooks, North Carolina Division of Archives and History, Raleigh

Alton Stewart Papers, in possession of Hal Stewart, Goldsboro, N.C., and Dave Stewart, Raleigh

William J. Tate Biographical File, National Air and Space Museum, Smithsonian Institution, Washington, D.C.

R. H. Whitfield Papers, North Carolina Collection, University of North Carolina Library at Chapel Hill

Wilbur and Orville Wright Papers, Library of Congress, Washington, D.C.

Forrest E. Wysong Biographical File, National Air and Space Museum, Smithsonian Institution, Washington, D.C.

NEWSPAPERS

*Asheville Citizen*

*Asheville Citizen-Times*

*Asheville Gazette-News*

*Austin (Tex.) Hornet*

*Beaufort-Hyde News*

*Beaufort News*

*Boston Christian Science Monitor*

*Carthage Moore County News*

*Chadbourn Journal*

*Chapel Hill Weekly*

*Charlotte Evening Chronicle*

*Charlotte News*

*Charlotte Observer*

*Clayton Chronicle*

Clayton News
Concord Times
Daily Charlotte Observer
Durham Herald-Sun
Durham Morning Herald
Durham Recorder
Eden Daily News
Elizabeth City Advance
Elizabeth City Independent
Elizabeth City Tar Heel
Fayetteville Observer
Fort Sill (Okla.) Trench and Camp
Goldsboro Headlight
Greensboro Daily News
Greensboro Patriot
Greensboro Record
Greensboro Telegram
Greenville Daily Reflector
Henderson French Broad Hustler
Kinston Free Press
Lexington Dispatch
Lumberton Robesonian
Manteo Coastlander
Manteo Coastland Times
Morganton Herald
New Bern Daily Journal
New York Evening Post
New York Herald Tribune
New York Times
New York World
Norfolk Gazette and Publick Ledger
Norfolk Journal
Norfolk Virginia Chronicle
Norfolk Virginian
Norfolk Virginian-Pilot
Oxford Torchlight
Philadelphia Evening Bulletin
Pinehurst Outlook
Pittsburgh Dispatch
Raleigh Caucasian
Raleigh Daily News

Raleigh Daily Tribune
Raleigh Farmer and Mechanic
Raleigh News and Observer
Raleigh Spectator
Raleigh State Chronicle
Raleigh Times
Reidsville Review
Reidsville Webster's Weekly
Rutland (Vt.) Daily Herald
Salisbury Carolina Watchman
Salisbury Evening Post
San Diego Tribune
San Diego Union
Sanford Carolina Watchman
Sanford Central Express
San Francisco Golden Era
San Francisco News Letter
Smithfield Herald
Southern Pines Pilot
Southern Pines Tourist
Spencer Crescent
Stars and Stripes (France)
Statesville Landmark
Statesville Sentinel
Stonewall Pamlico Enterprise
Wake Forest College Old Gold and Black
Warrenton Gazette
Warrenton Warren Record
Washington (D.C.) Evening Critic
Washington (D.C.) Evening Star
Washington (D.C.) Forney's Sunday
    Chronicle
Washington (D.C.) National Republican
Washington (D.C.) Post
Weldon Roanoke News
Wilkesboro Chronicle
Wilmington Daily Journal
Wilmington Evening Dispatch
Wilmington Messenger
Wilmington Morning Herald
Wilmington Morning Star

Wilmington Morning Sun
Wilson Advance
Wilson Daily Times

Winston-Salem Journal
Winston-Salem Journal and Sentinel
Winston-Salem Union Republican

BOOKS

*Aircraft Yearbook Issued by Manufacturers Aircraft Association, Inc.* New York: Doubleday, Page, 1920.

Albertson, Catherine. *Wings over Kill Devil and Legends of the Dunes.* Elizabeth City, N.C.: Privately Printed, 1928.

Allott, Kenneth. *Jules Verne.* New York: Macmillan, 1941.

*Annual Report of the Commissioner of Patents.* Washington, D.C.: U.S. Government Printing Office, 1905, 1910.

Bates, Jo Anne Heath, ed. *The Heritage of Currituck County, North Carolina.* Winston-Salem: Hunter Publishing Co., 1985.

Bates, Joseph Clemont, ed. *History of the Bench and Bar of California.* San Francisco: Bench and Bar Publishing Co., 1886.

Becker, Beril. *Dreams and Realities of the Conquest of the Skies.* New York: Atheneum, 1967.

Bensen, Igor, with Paul Bergen Abbott. *A Dream of Flight.* Indianapolis: Abbott Co., 1992.

Bishir, Catherine W., Charlotte V. Brown, Carl R. Lounsbury, and Ernest H. Wood III. *Architects and Builders in North Carolina: A History of the Practice of Builders in North Carolina.* Chapel Hill: University of North Carolina Press, 1990.

Brooks-Pazmany, Kathleen. *United States Women in Aviation, 1919–1929.* Washington, D.C.: Smithsonian Institution Press, 1993.

Brown, Aycock. *The Birth of Aviation, Kitty Hawk, N.C.* Winston-Salem: Collins Press, 1953.

Brown, Walter, Jr., ed. *An American for Lafayette: The Diaries of E. C. C. Genet, Lafayette Escadrille.* Charlottesville: University Press of Virginia, 1981.

Chambers, Lenoir, Joseph E. Shank, and Harold Sugg. *Salt Water and Printer's Ink: Norfolk and Its Newspapers, 1865–1965.* Chapel Hill: University of North Carolina Press, 1967.

Clark, Chatham. *The Wayward Balloon, a True Story.* Elizabethtown, N.C.: Hill House Press, 1979.

Clark, Sidney A. *Cathedral France.* New York: Robert M. McBride & Co., 1931.

Corn, Joseph J. *The Winged Gospel: America's Romance with Aviation, 1900–1950.* New York: Oxford University Press, 1983.

Cotten, Jerry W. *Light and Air: The Photography of Bayard Wootten.* Chapel Hill: University of North Carolina Press, 1998.

Crouch, Tom. *The Bishop's Boys: A Life of Wilbur and Orville Wright*. New York: Norton, 1989.

Daniels, Josephus. *Our Navy at War*. Washington, D.C.: Pictorial Bureau, 1922.

——. *The Wilson Era*. 2 vols. Chapel Hill: University of North Carolina Press, 1944.

Davis, Pat Dula, and Kathleen Hill Hamilton, eds. *Historic Carteret County, North Carolina*. 2 vols. Published by the Carteret Historical Research Association, Beaufort, 1975, in cooperation with Hunter Publishing Co., Winston-Salem.

*Dell School Catalogue, 1908–1909*. N.p., 1908.

Dudley, Jack. *Carteret Waterfowl Heritage*. Burtonsville, Md.: Decoy Magazine, 1993.

Dula, W. C. *Durham and Her People*. Durham, N.C.: N.p., 1951.

Falk, Peter H., ed. *Who Was Who in American Art*. Guilford, Conn.: Sound View Press, 1985.

Frost, Robert. *Kitty Hawk, 1894*. New York: Henry Holt and Co., 1962.

Gibbs-Smith, Charles H. *Sir George Cayley's Aeronautics, 1796–1855*. London: Her Majesty's Stationery Office, 1962.

Gillispie, Charles C. *The Montgolfier Brothers and the Invention of Aviation, 1783–1784*. Princeton: Princeton University Press, 1983.

Glines, Carroll V. *Roscoe Turner: Aviation's Master Showman*. Washington, D.C.: Smithsonian Institution Press, 1995.

Goldstein, Laurence. *The Flying Machine and Modern Literature*. Houndsmills, N.H.: Macmillan, 1986.

Hall, Lewis P. *Land of the Golden River: Historical Events and Stories of Southeastern North Carolina and the Lower Cape Fear*. Wilmington: N.p., 1975.

——. *Old Wilmington and the Greater in Its March to the Sea*. Wilmington: N.p., 1975.

Harmon, Nolan B. *The Famous Case of Myra Clark Gaines*. Baton Rouge: Louisiana State University Press, 1946.

Haynes, Richard M. *The Wright Brothers*. Englewood Cliffs, N.J.: Silver-Burdett, 1971.

Hill, Mrs. Fred. *Historic Carteret County, North Carolina*. Beaufort: Carteret Historical Research Association, 1975.

Ireland, Robert E. *Entering the Auto Age: The Early Automobile in North Carolina, 1900–1930*. Raleigh: Historical Publications Section, North Carolina Division of Archives and History, 1990.

Jacobs, David M. *The UFO Controversy in America*. Bloomington: Indiana University Press, 1975.

Jakab, Peter L., and Rick Young, eds. *The Published Writings of Wilbur and Orville Wright*. Washington, D.C.: Smithsonian Institution Press, 2000.

Jane. *Jane's All the World's Aircraft*. London: Sampson Low, Marston and Co., Ltd., 1912, 1922.

Johnson, F. Roy, and E. Frank Stephenson Jr. *The Gatling Gun and Flying Machine of Richard and Henry Gatling*. Murfreesboro, N.C.: Johnson Publishing Co., 1979.

Karolevitz, Robert F., and Ross S. Fenn. *Flight of Eagles: The Story of the American Kosciuszko Squadron in the Polish-Russian War, 1919–1920*. Sioux Falls, S.D.: Brevet Press, Inc., 1974.

Keever, Homer M. *Iredell: Piedmont County*. N.p.: Brady Printing Co., 1976.

Kelly, Fred C., ed. *Miracle at Kitty Hawk: The Letters of Wilbur and Orville Wright*. New York: Farrar, Straus and Young, 1959; Da Capo Press, 1996.

Lefler, Hugh T., ed. *A New Voyage to Carolina by John Lawson*. Chapel Hill: University of North Carolina Press, 1957.

Lefler, Hugh T., and Albert Ray Newsome. *North Carolina: The History of a Southern State*. Chapel Hill: University of North Carolina Press, 1973.

Levine, Isaac Don. *Mitchell: Pioneer of Air Power*. New York: Duell, Sloan and Pearce, 1958.

Liberatore, E. K. *Helicopters before Helicopters*. Malabar, Fla.: Krieger Publishing Co., 1998.

Longstreet, Stephen. *The Canvas Falcons: The Men and the Planes of World War I*. New York: Barnes and Noble Books, 1970.

Mackworth-Praed, Ben. *Aviation: The Pioneer Years*. Secaucus, N.J.: Chartwell Books, 1990.

MacNeill, Ben Dixon. *The Hatterasman*. Winston-Salem: John F. Blair, 1958.

Mansfield, Howard. *Skylark: The Life, Lies, and Inventions of Harry Atwood*. Hanover, N.H.: University Press of New England, 1999.

Mason, Francis K., and Martin C. Windrow, comps. *Air Facts and Feats: A Record of Aerospace Achievements*. Garden City, N.Y.: Doubleday, 1970.

Massengill, Stephen E. *Around Southern Pines: A Sandhills Album*. Photographs by E. C. Eddy. Dover, N.H.: Chalford Publishing Corp., 1998.

McAdoo, Donald, and Carol McAdoo. *Reflections on the Outer Banks*. Manteo: Island Publishing House, 1976.

McConnell, James R. *Flying for France with the American Escadrille at Verdun*. New York: Doubleday, Page, 1917.

McFarland, Marvin W., ed. *Papers of Wilbur and Orville Wright: Including the Wright-Chanute Letters and Other Papers of Octave Chanute*. 2 vols. New York: McGraw-Hill, 1953.

Murray, Elizabeth. *Wake: Capital County of North Carolina: Prehistory through Centennial*. Raleigh: Capital County Publishing Co., 1983.

Norman, Aaron. *The Great Air War: The Men, the Planes, the Saga of Military Aviation: 1914–1918*. New York: Macmillan, 1968.

*Official Gazette of the United States Patent Office*. Washington, D.C.: U.S. Government Printing Office, 1910, 1912, 1914, 1922.

Parker, J. Mayon. *The Ahoskie Era of Hertford County*. Ahoskie, N.C.: Parker Brothers, Inc., 1939.

Parsons, Edwin C. *The Great Adventure*. Garden City, N.Y.: Doubleday, Doran, 1937.

——. *I Flew with the Lafayette Escadrille*. Indianapolis: E. C. Seale, 1953.

Powell, William S. *North Carolina Fiction, 1734–1957: An Annotated Bibliography*. Chapel Hill: University of North Carolina Press, 1958.

——. *The North Carolina Gazetteer*. Chapel Hill: University of North Carolina Press, 1968.

——, ed. *Dictionary of North Carolina Biography*. 6 vols. Chapel Hill: University of North Carolina Press, 1979–96.

Power, J. Tracy. *Lee's Miserables: Life in the Army of Northern Virginia from the Wilderness to Appomattox*. Chapel Hill: University of North Carolina Press, 1998.

Reeves, Bill. *Southport (Smithwick): A Chronology*. Wilmington: New Hanover County Public Library, n.d.

Reilly, Thomas. *Jannus: An American Flier*. Gainesville, Fla.: University Press of Florida, 1997.

Renstrom, Arthur G. *Wilbur and Orville Wright: A Chronology Commemorating the One Hundredth Anniversary of the Birth of Orville Wright*. Washington, D.C.: Library of Congress, 1997.

Robie, Bill. *For the Greatest Achievement: A History of the Aero Club of America and the National Aeronautic Association*. Washington, D.C.: Smithsonian Institution Press, 1993.

Rockwell, Paul A. *American Fighters in the Foreign Legion, 1914–1918*. Garden City, N.Y.: Country Life Press, 1925.

Rogers, Dennis. *Home Grown*. Raleigh: News and Observer Publishing Co., 1979.

Rowe, Nellie M. *Discovering North Carolina*. Chapel Hill: University of North Carolina Press, 1940.

Schofel, Johann W. A. *Hirum-Harum: Ein Satirsh-komisscher Original-Roman*. Nuremberg: Schneider und Weigel, 1789.

Scott, Phil. *The Pioneers of Flight: A Documentary History*. Princeton: Princeton University Press, 1999.

——. *Shoulders of Giants: A History of Human Flight to 1919*. Reading, Mass.: Wesley-Addison Publishing Co., 1995.

Shamburger, Page, and Joe Christy. *Command the Horizon: A Pictorial History of Aviation*. New York: Barnes, 1968.

Shanks, Ralph, Wick York, and Lisa Woo Shanks. *The U.S. Life-Saving Service: Heroes, Rescue, and Architecture*. Petaluma, Calif.: Costano Books, 1996.

Smith, Richard K. *First Across!: The U.S. Navy's Transatlantic Flight of 1919*. Annapolis, Md.: Naval Institute Press, 1973.

Tate, William J. *Wings over Kill Devil: Brochure of the Twenty-fifth Anniversary of the First Successful Airplane Flight, 1903–1928*. N.p.: Published by the author, 1928.

Tillman, Stephen F. *Man Unafraid: The Miracle of Military Aviation*. Washington, D.C.: Army Times Publishing Co., 1958.

Trimble, William F. *High Frontier: A History of Aeronautics in Pennsylvania*. Pittsburgh: University of Pittsburgh Press, 1992.

Vecsey, George, and George C. Dade. *Getting off the Ground: The Pioneers of Aviation Speak for Themselves*. New York: Dutton, 1979.

Verne, Jules. *Maitre du Monde*. Paris: Collection Hetzel, n.d.

———. *Master of the World*. London: S. L. Marston and Co., n.d. [1904].

Villard, Henry Serrano. *Contact: The Story of the Early Birds*. New York: Bonanza Books, 1953.

Walser, Richard. *Literary North Carolina: A Brief Survey*. Raleigh: Historical Publications, North Carolina Division of Archives and History, 1970.

Walsh, John Evangelist. *One Day at Kitty Hawk: The Untold Story of the Wright Brothers and the Airplane*. New York: Harcourt, Brace, 1975.

West, John Foster. *This Proud Land: The Blue Ridge Mountains*. Charlotte: McNally and Loftin, 1974.

Whitehouse, Arch. *The Early Birds: The Wonders and Heroics of the First Decade of Flight*. Garden City, N.Y.: Doubleday, 1965.

*Who Was Who in America, 1847–1892*. Chicago: Marquis Who's Who, 4th printing, 1960.

Wohl, Robert. *A Passion for Wings: Aviation and the Western Imagination, 1908–1918*. New Haven: Yale University Press, 1994.

Wolfe, Thomas. *Look Homeward, Angel: A Story of the Buried Life*. New York: Simon and Schuster, 1929.

ARTICLES

"Airship Inventor Offered to Carry the Mail Sixty Years Ago." *Popular Mechanics*, June 1910, 837.

"Army Airplane Receipts, 1908–1995." *Aerospace Historian* 13 (Winter 1966): 164.

Arthur, Billy. "The Flying Parson of Sampson County." *The State*, February 1988, 12–15.

———. "The World's First Airplane Casualty." *The State*, December 1990, 11.

Beach, Stanley Yale. "The New York Aero Show." *Scientific American* 106 (May 25, 1912): 483.

Bellefleur, John R. "The Last Early Aeroplane? An Essay on an Old Family

Legend." *WW I Aero: The Journal of the Early Aeroplane* 146 (November 1994): 327–48.

Bonney, Walter T. "Chiefs of the Army Air Force." *Air Power Historian* 7 (July 1960): 88–100.

Bowers, Ray L. "Aviation on Trial: Part II." *Air Power Historian* 8 (April 1961): 88–100.

Brown, Aycock. "All in One Lifetime (A Tribute to Luther Paul)." Typescript copy of undated article in the *Beaufort News* shortly after Paul's death.

Buffington, H. Glenn. "Flying Life of Viola Gentry." *AAHS Journal* 13 (Summer 1968): 125–26.

Burnelli, Vincent J., with Booton Herndon. "The Wackiest Plane in World War I." *True: The Man's Magazine*, pp. 59, 72–79. Undated clipping, Christmas Biographical File, National Air and Space Museum, Smithsonian Institution, Washington, D.C.

"Captain William J. Tate." *U.S. Air Services*, July 8, 1953, 6.

"Double Transcontinental Aerial Derby." *U.S. Air Services*, November 2, 1969, 6–11.

Drinkwater, Alpheus W. "I Knew Those Wright Brothers Were Crazy." *Reader's Digest*, November 1956, 283–94.

Evans, Charles Morgan. "Steam-Powered Pioneer." *Aviation*, May 1993, 48–53, 71–72.

"Forrest Wysong: Our New President." *Chirp* 83 (January 1985): 30–34.

Foulois, Benjamin D. "Early Flying Experiences." *Air Power Historian* 3 (Winter 1956): 127.

Frost, Robert. "A Trip to Currituck, Elizabeth City, and Kitty Hawk (1894)." *North Carolina Folklore* 16 (May 1968): 3–8.

"The Great Balloon Hoax." *The State*, September 2, 1961, 9–10.

Gresham, N. E. "Editor's Portfolio." *Wake Forest Student* 36 (December 1919): 40.

Hall, Lewis P. "The American Aeroplane Company of Wilmington." *Old Wilmington and the Greater in Its March to the Sea*, 296–97. Wilmington: N.p., 1975.

———. "Another First Flight." *The State*, August 10, 1968, 9.

———. "William D. Polite, Head-Waiter Extraordinary." *Land of the Golden River: Historical Events and Stories of Southeastern North Carolina and the Lower Cape Fear*, 99–101. Wilmington: N.p., 1975.

Hayman, L. D. "The Hayman Clan in North Carolina, 1818–1950." N.p., n.d. This is a fifteen-page genealogy issued in photocopy.

Hufham, Joseph S. "Who Carried Away the UFO at Delco?" *The State*, February 1978, 23–24.

"Kitty Hawk Twenty-five Years after the Event." *U.S. Air Services*, January 1919, 28–31.

"Landing Fields in the United States." *Aerial Age* 15 (May 1922): 22.

Lowe, Thomas E. "Johnny Crowell: Aircraftsman, Exhibition Pilot, and Inventor." *AAHS Journal* 27 (Fall 1982): 210–15.

Marden, Luis. "She Wore the World's First Wings." *Outer Banks* (1984): 25–29.

Mattox, Henry E. "Chariots of Wrath." *North Carolina Historical Review* 73 (July 1996): 297–308.

Maurer, Maurer. "The First Aero Squadron, 1913–1917." *Air Power Historian* 4 (October 1957): 207–15.

"The Men and the Machines: Part I." *Air Power Historian* 3 (October 1956): 171–80.

Moore, Harry P. "Reporter Who 'Scooped' the World on First Flight Meets Wright." In Shank's Raw Materials, vol. 4, Shank Papers.

"Mortality in Army Aviation." *AAHS Journal* 9 (Summer 1964): 113.

Moscowitz, Sam. "Science Fiction of William Henry Rhodes." Introduction in Daniel O'Connell, ed., *Caxton's Book: A Collection of Essays, Poems, Tales, and Sketches*. Westport, Conn.: Hyperion Press, 1974, iv–v.

Norris, Mrs. Garland. "The Flying Parson." *The State*, September 29, 1934.

Olds, Fred A. "The Mother of the Airplane." *Orphan's Friend and Masonic Journal* 43 (September 20, 1918): 1.

Owen, Guy. "Robert Frost at Kitty Hawk." *The State*, March 1975, 14–15.

Parramore, Thomas C. "The Flight of the 'Bumble Bee': The First Helicopter?" *WWI Aero: The Journal of the Early Aeroplane* 158 (November 1997): 8–12.

———. "The North Carolina Background of Richard Jordan Gatling." *North Carolina Historical Review* 41 (January 1964): 54–61.

Paul, Daphne. "Accomplishments of William Luther Paul, 1869–1946." *Historic Carteret County, North Carolina*. 2 vols. Published by the Carteret Historical Association, Beaufort, 1975, 1:87–88.

Paul, Grayden. "Grayden Paul Reminisces about His Boyhood Days in Davis Shore." In Pat Dula Davis and Kathleen Hill Hamilton, eds., *The Heritage of Carteret County, North Carolina*, 2:66–71. 2 vols. Published by the Carteret Historical Research Association, Beaufort, 1975, in cooperation with Hunter Publishing Co., Winston-Salem.

Pierce, T. H. "First to Jump." *The State*, January 1975, 8–10.

Rockwell, Paul A. "Writings of the American Pilots in the Escadrille Lafayette." *Ex Libris* 1 (November 1923): 131–35.

Rogers, Frances. "Luther Paul: Father of Beaufort Auto Dealer, Was County's Pioneer Inventor." Undated clipping from *Beaufort News*, furnished to author by Prof. Charles Paul.

Selby, Marjorie, and R. S. Spencer Jr. "Fairfield." In *Hyde County History: A Hyde County Bicentennial Project*, 37–38. Charlotte: Herb Eaton, Inc., 1976.

Sharpe, Bill. "Johnny May Go to Washington." News release, State News Bureau, Raleigh, November 1948, McSurely Papers.

Shields, Mattie H. "Jacob Aaron Hill." In Wade Duncan, ed., *The Heritage of Stokes County*, 1:241–42. 2 vols. N.p., 1990.

"Sketch of the Late William H. Rhodes," by "Scintilla Juris." *Bench and Bar*. Excerpt furnished to author by Michael Rosen, San Francisco, from an otherwise unidentified volume.

Stearns, Myron M. "The Flying Wing." *Popular Mechanics*, August 1929, 234–39.

Tate, Daniel Grady. "W. J. Tate and Family." In Jo Anne Heath Bates, ed., *The Heritage of Currituck County, North Carolina*, 2:419–20. 2 vols. Winston-Salem: Hunter Publishing Co., 1985.

Tate, William J. "I Was Host to the Wright Brothers at Kitty Hawk." *U.S. Air Services*, December 1943, 29–30, 189–92.

———. "A Lighthouse Keeper's Connection with Pioneers of Aviation." *Lighthouse Service Bulletin*, January 1929, 272–73.

———. "With the Wrights at Kill Devil Hill." *Aeronautic Review* 6 (December 1928): 189–92.

"Toronto–New York Handicap Race." *U.S. Air Services*, September 21, 1919, 20.

Van Vliet, John D. "The Inside Story of the Bullet." *Popular Aviation*, July 1934, 21–22, 37, 50–52, 60.

Walser, Richard. "Jules Verne's Fantastic Voyages to North Carolina." *The State*, December 1987, 32–33.

Watson, Alan D. "Sailing under Steam: The Advent of Steam Navigation in North Carolina to the Civil War." *North Carolina Historical Review* 75 (January 1998): 31–40.

*Who's Who in Aviation, 1942–1943*. Writers' Program of the Work Project Administration. Chicago: Ziff-Davis Publishing Co., 1942.

Winters, S. R. "Life-Saving Airplanes." *Scientific American* 137 (December 1927): 499.